T0176982

BIOMEDICAL DEVICES

BIOMEDICAL DEVICES

Design, Prototyping, and Manufacturing

Edited By

TUĞRUL ÖZEL
PAOLO JORGE BÁRTOLO
ELISABETTA CERETTI
JOAQUIM DE CIURANA GAY
CIRO ANGEL RODRIGUEZ
JORGE VICENTE LOPES DA SILVA

Published by John Wiley & Sons, Inc., Hoboken, New Jersey
Published simultaneously in Canada

For general information on our other products and services or for technical support, please contact our Customer Care Department within the United States at (800) 762-2974, outside the United States at (317) 572-3993 or fax (317) 572-4002.

Wiley also publishes its books in a variety of electronic formats. Some content that appears in print may not be available in electronic formats. For more information about Wiley products, visit our web site at www.wiley.com.

Library of Congress Cataloging-in-Publication Data:

Names: Özel, Tuğrul, 1967- editor. | Bártolo, Paulo, editor. | Ceretti,
 Elisabetta, 1966- editor. | Ciurana Gay, Joaquim De, editor. | Rodriguez,
 Ciro Angel, 1967- editor. | Silva, Jorge Vicente Lopes da, 1963- editor.
Title: Biomedical devices : design, prototyping, and manufacturing / edited
 by Tuğrul Özel, Paolo Jorge Bártolo, Elisabetta Ceretti, Joaquim De Ciurana Gay,
Ciro Angel Rodriguez, and Jorge Vicente Lopes Da Silva.
Description: Hoboken, New Jersey : John Wiley & Sons, 2016. | Includes
 bibliographical references and index.
Identifiers: LCCN 2016018375| ISBN 9781118478929 (cloth) | ISBN 9781119267041
 (epub)
Subjects: LCSH: Medical electronics–Design and construction.
Classification: LCC R856 .B487 2016 | DDC 610.28/4–dc23 LC record available at
 https://lccn.loc.gov/2016018375

Cover image courtesy of the Editors.

Typeset in 10/12pt TimesLTStd by SPi Global, Chennai, India

Printed in the United States of America

10 9 8 7 6 5 4 3 2 1

CONTENTS

5   Machining Applications   99

Tuğrul Özel, Elisabetta Ceretti, Thanongsak Thepsonthi,
and Aldo Attanasio

6   Inkjet- and Extrusion-Based Technologies   121

Karla Monroy, Lidia Serenó, Joaquim De Ciurana Gay, Paulo Jorge Bártolo,
Jorge Vicente Lopes Da Silva, and Marco Domingos

7   Certification for Medical Devices   161

Corrado Paganelli, Marino Bindi, Laura Laffranchi, Domenico Dalessandri,
Stefano Salgarello, Antonio Fiorentino, Giuseppe Vatri,
and Arne Hensten

CONTRIBUTORS

Aldo Attanasio, Department of Mechanical Engineering and Industrial Engineering, University of Brescia, Brescia, Lombardy, Italy

Paulo Jorge Bártolo, School of Mechanical, Aerospace and Civil Engineering, Manchester Institute of Biotechnology, University of Manchester, Manchester, UK

Karen Baylón, Department of Mechanical Engineering, Instituto Tecnológico y de Estudios Superiores de Monterrey, Campus Monterrey, Monterrey, Nuevo León, Mexico

Marino Bindi, Department of Medical and Surgical Specialties, Radiological Sciences, and Public Health, University of Brescia, Brescia, Lombardy, Italy

Elisabetta Ceretti, Department of Mechanical and Industrial Engineering, University of Brescia, Brescia, Lombardy, Italy

Luis Criales, Department of Industrial and Systems Engineering, School of Engineering, Rutgers University, Piscataway, NJ, USA

Jorge Vicente Lopes Da Silva, Technological Center Renato Archer, Centro de Tecnologia da Informação Renato Archer, Brazilian Ministry of Science and Technology, Campinas, São Paulo, Brazil

Domenico Dalessandri, Department of Medical and Surgical Specialties, Radiological Sciences, and Public Health, University of Brescia, Brescia, Lombardy, Italy

Joaquim De Ciurana Gay, Department of Mechanical Engineering and Industrial Construction, University of Girona, Girona, Catalonia, Spain

Marco Domingos, School of Mechanical, Aerospace and Civil Engineering, Manchester Institute of Biotechnology, University of Manchester, Manchester, UK

Alex Elias-Zuñiga, Department of Mechanical Engineering, Instituto Tecnológico y de Estudios Superiores de Monterrey, Campus Monterrey, Monterrey, Nuevo León, Mexico

Inés Ferrer, Department of Mechanical Engineering and Industrial Construction, University of Girona, Girona, Catalonia, Spain

Antonio Fiorentino, Department of Mechanical and Industrial Engineering, University of Brescia, Brescia, Lombardy, Italy

Maria Luisa Garcia-Romeu, Department of Mechanical Engineering and Industrial Construction, University of Girona, Girona, Catalonia, Spain

Claudio Giardini, Department of Mechanical and Industrial Engineering, University of Brescia, Brescia, Lombardy, Italy

Jordi Grabalosa, Department of Mechanical Engineering and Industrial Construction, University of Girona, Girona, Catalonia, Spain

Arne Hensten, Faculty of Health Sciences, UiT The Arctic University of Norway, Tromsø, Norway

Laura Laffranchi, Department of Medical and Surgical Specialties, Radiological Sciences, and Public Health, University of Brescia, Brescia, Lombardy, Italy

Karla Monroy, Department of Mechanical Engineering and Industrial Construction, University of Girona, Girona, Catalonia, Spain

Tuğrul Özel, Department of Industrial and Systems Engineering, School of Engineering, Rutgers University, Piscataway, NJ, USA

Corrado Paganelli, Department of Medical and Surgical Specialties, Radiological Sciences, and Public Health, University of Brescia, Brescia, Lombardy, Italy

Ciro Angel Rodriguez, Department of Mechanical Engineering, Instituto Tecnológico y de Estudios Superiores de Monterrey, Campus Monterrey, Monterrey, Nuevo León, Mexico

Stefano Salgarello, Department of Medical and Surgical Specialties, Radiological Sciences, and Public Health, University of Brescia, Brescia, Lombardy, Italy

Lidia Serenó, Department of Mechanical Engineering and Industrial Construction, University of Girona, Girona, Catalonia, Spain

Daniel Teixidor Ezpeleta, Department of Mechanical Engineering and Industrial Construction, University of Girona, Girona, Catalonia, Spain

Thanongsak Thepsonthi, Department of Industrial Engineering, Burapha University, Chon Buri, Thailand

Giuseppe Vatri, Major Prodotti Dentari Spa, Moncalieri, TO, Italy

FOREWORD

There has been an increasing demand for biomedical devices including various instruments, apparatuses, implants, *in vitro* reagents, and similar articles to diagnose, prevent, or treat diseases, improve human health, prolong human life, and recover from serious injuries. Biomedical devices that we utilize are not only continuously getting smaller and more effective but are also designed with more customized functionalities. In response to that demand, new design innovations, new materials, and prototypes of novel medical devices have been introduced on a regular basis. Design, prototyping, and manufacturing techniques for these materials and designs have also been continuously developing in parallel to the needs in biomedical device demands. There have been a large number of books written about design and manufacturing of various products but biomedical device manufacturing remains less covered than other well-known microelectronics and consumer products.

This book brings authors from institutions around the world, perhaps one of the few wide-ranging books on manufacturing processes for medical devices with coverage of various materials including metals and polymers. The book aims to reach audiences such as practicing engineers who are working in medical device industry, students in the biomedical device manufacturing courses, and faculty/researchers who are conducting research in medical device design, prototyping, and manufacturing.

Chapter 1, written by Joaquim De Ciurana, Tuğrul Özel, and Lidia Sereró, provides an introduction with classification and regulation specification about medical devices with sample manufacturing processes and applications. Chapter 2, prepared by Inés Ferrer Real, Jordi Grabalosa, Alex Elias-Zuniga, and Ciro Rodriguez, summarizes design issues and methodologies applicable to well-known medical device prototypes. Chapter 3, prepared by Karen Baylón, Elisabetta Ceretti, Claudio Giardini, and M. Luisa Garcia-Romeu, describes the issues in modeling and analysis for forming processes along with the comparison of different modeling approaches. Chapter 4, prepared by Tuğrul Özel, Joaquim De Ciurana, Daniel Teixidor, and Luis Criales, describes some of the manufacturing processes based on laser processing with several examples for medical devices. Chapter 5, prepared by Tuğrul Özel, Elisabetta Ceretti, Thanongsak Thepsonthi, and Aldo Attanasio, discusses machining applications and manufacturing processes to be used for medical devices made out of metals and plastics. Extrusion- and inkjet-based processes and applications are presented in Chapter 6, which was prepared by Karla Monroy, Lidia Serenó, Joaquim De Ciurana, Paulo Jorge Bártolo, Jorge Da Silva, Marco Domingos, Corrado Paganelli, Marino Bindi, Laura Laffranchi, Domenico Dalessandri, Stefano Salgarello, Antonio Fiorentino, Giuseppe Vatri, and Arne Hensten prepared the Chapter 7 on certification of medical devices.

We thank all the authors who contributed to this book. We also extend our thanks to Ms Anita Lekhwani and Ms Sumathi Elangovan of John Wiley who assisted us in all stages of preparing this book for the publication.

T. ÖZEL, P. BÁRTOLO, E. CERETTI, J. CIURANA GAY,
C. A. RODRÍGUEZ, J. V. LOPES DA SILVA
June 2016

1

OVERVIEW

JOAQUIM DE CIURANA GAY

Department of Mechanical Engineering and Industrial Construction, University of Girona, Girona, Catalonia, Spain

TUĞRUL ÖZEL

Department of Industrial and Systems Engineering, School of Engineering, Rutgers University, Piscataway, NJ, USA

LIDIA SERENÓ

Department of Mechanical Engineering and Industrial Construction, University of Girona, Girona, Catalonia, Spain

1.1 INTRODUCTION

Medical devices are defined as articles that are intended to be used for medical purposes. Several official definitions exist for the term "medical device" depending on the geographic market. Therefore, medical device definition, classification, and regulation follow market location and governmental regulations according to the required level of control considering invasiveness, contact to the patient, and potential risk in case of misuse or failure. This situation concerning the differences in classification strategies has blocked the spread of innovative medical devices across countries. Nevertheless, in 2011, the International Medical Device Regulators Forum (IMDRF)

Biomedical Devices: Design, Prototyping, and Manufacturing, First Edition.
Edited by Tuğrul Özel, Paolo Jorge Bártolo, Elisabetta Ceretti, Joaquim De Ciurana Gay, Ciro Angel Rodriguez, and Jorge Vicente Lopes Da Silva.

was conceived to discuss future directions to harmonize the medical device regulatory field and accelerate international convergence.

Two of the most important medical device markets worldwide are the European and the North American. Therefore, the official definitions and classifications of both regions are detailed in this chapter.

For the European market, medical devices are governed by a regulatory framework of three directives:

- 93/42/EEC: Medical Devices Directive (MDD)
- 90/385/EEC: Active Implantable Medical Device Directive (AIMDD)
- 98/79/EC: *In vitro* diagnostic medical devices (IVDMD)

According to them, a medical device is defined as "any instrument, apparatus, appliance, software, material or other article, whether used alone or in combination, including the software intended by its manufacturer to be used specifically for diagnostic and/or therapeutic purposes and necessary for its proper application, intended by the manufacturer to be used for human beings for the purpose of:

- diagnosis, prevention, monitoring, treatment or alleviation of disease,
- diagnosis, monitoring, treatment, alleviation of or compensation for an injury or handicap,
- investigation, replacement or modification of the anatomy or of a physiological process,
- control of conception,

and which does not achieve its principal intended action in or on the human body by pharmacological, immunological or metabolic means, but which may be assisted in its function by such means."

Similarly, for the North American market and as a part of the Federal Food Drug and Cosmetic Act (FDC Act), a medical device is defined as "an instrument, apparatus, implement, machine, contrivance, implant, *in vitro* reagent, or other similar or related article, including a component part, or accessory which is:

- recognized in the official National Formulary, or the United States Pharmacopeia, or any supplement to them,
- intended for use in the diagnosis of disease or other conditions, or in the cure, mitigation, treatment, or prevention of disease, in man or other animals, or
- intended to affect the structure or any function of the body of man or other animals, and which does not achieve its primary intended purposes through chemical action within or on the body of man or other animals and which is not dependent on being metabolized for the achievement of any of its primary intended purposes."

Therefore, any product labeled, promoted, or used in a manner that meets the above-mentioned definition will be regulated by the US Food and Drug

Administration (US FDA) as a medical device and will be subjected to pre- and postmarketing regulatory controls.

Both definitions of medical devices exclude other regulated products such as drugs, the primary intended use of which is achieved through chemical action or by being metabolized by the body, biological products including blood and blood products, or products used with animals.

In order to classify a medical device, the manufacturer should, first of all, decide whether the product concerned is considered a medical device as defined in the previous section. Then, depending on the situation, medical devices can be classified following national or governmental rules. In this chapter, the classification given is based on the European Union (EU) and the US FDA regulations.

According to the EU, the classification of medical devices is based on the potential risks associated with the devices. This approach allows the use of a set of criteria that can be combined in various ways and be applied to a vast range of different medical devices and technologies. These criteria are referred to as the "classification rules" and are described in Annex IX of Directive 93/42/EEC. Therefore, the medical device manufacturer must determine the type of device following the rules listed in Annex IX. The rules depend on a series of factors including

- *duration:* how long the device is intended to be in continuous use,
- *invasiveness:* whether or not the device is invasive or surgically invasive,
- *type:* whether the device is implantable or active,
- *function:* whether or not the device contains a substance, which in its own right is considered to be a medicinal substance and has action ancillary to that of the device,

and are divided as follows:

- *Rules 1–4:* for noninvasive devices,
- *Rules 5–8:* for invasive devices,
- *Rules 9–12:* for active devices,
- *Rules 13–18:* special rules for products that merit a higher classification than they might otherwise be assigned.

When multiple rules apply, the manufacturer must use the highest risk class. Nevertheless, a small number of products may be more difficult to classify due to their unusual nature or situations where the classification would result in the wrong level of conformity assessment in light of the hazard represented by the devices.

Furthermore, based on these rules described in Directive 93/42/EEC, the devices are divided into four classes, ranging from low risk to high risk:

- *Class I:* low-risk medical devices,
- *Class IIa:* medium-risk medical devices,

- *Class IIb:* medium-risk medical devices,
- *Class III:* high-risk medical devices.

Thus, in order to classify a medical device, the manufacturer must determine the classification of the medical device (class I, class IIa, class IIb, or class III) considering the Annex IX rules described later. Then, a notified body has to carry out the appropriate conformity assessment procedure to validate and confirm the classification.

As an example, a manufacturer willing to classify a silicone tracheal stent must consider the rules associated with an invasive medical device (Rules 5–8):

- **Rule 5**   *(invasive in body orifice or stoma—not surgically)*

If it is for transient use	Class I
If it is for short-term use	Class IIa
However, if it is for oral cavity, ear canal, or nasal cavity	Class I
If it is for long-term use	Class IIb
However, if it is for oral cavity, ear canal, or nasal cavity and it will not be absorbed by the mucous membrane	Class IIa
If it is connected to an active medical device in class IIa or higher	Class IIa

- **Rule 6**   *(surgically invasive—transient use)*

If it is surgically invasive for transient use	Class IIa
If it is used to control/diagnose/monitor/correct a defect of the heart or the central circulatory system through direct contact	Class III
If it is used for the central nervous system (direct contact)	Class III
If it is a reusable surgical instrument	Class I
If it is used to supply energy or ionizing radiation	Class IIb
If it has a biological effect (mainly or wholly absorbed)	Class IIb
If it is intended to administer medicines in a potentially hazardous manner	Class IIb

- **Rule 7**   *(surgically invasive—short-term use)*

If it is surgically invasive for short-term use	Class IIa
If it is used to control/diagnose/monitor/correct a defect of the heart or the central circulatory system through direct contact	Class III
If it is used for the central nervous system (direct contact)	Class III
If it is used to supply energy or ionizing radiation	Class IIb
If it has a biological effect (mainly absorbed)	Class III
If it undergoes chemical changes in the body, or if it administers medicines (not in teeth)	Class IIb

- **Rule 8**   *(surgically invasive—long-term use or implantable devices)*

If it is surgically invasive for long-term use or if it is an implantable device	Class IIb
If it has to be placed in teeth	Class IIa
If it has to be in contact with the heart or central circulatory/nervous system	Class III
If it has a biological effect (or mainly absorbed)	Class III
If it undergoes chemical changes in the body, or if it administers medicines (not in teeth)	Class III
For specific derogation: breast implants, hip, knee, and shoulder joint replacements	Class III

Specifically for the example, the manufacturer must consider the following:

- *Duration:* the silicone stent will be placed inside the trachea for more than 30 day; therefore, the device is for long-term use (**Rule 8**).
- *Invasiveness:* the stent will be totally introduced inside the orifice of the trachea using a bronchoscope and anesthesia (surgical operation); therefore, the device is considered an implantable device (**Rule 8**).

Taking these considerations into account, a simple silicone tracheal stent must be considered a class IIb medical device, because it is a long-term implantable device not placed in teeth, without contact with the circulatory or nervous system, without a biological effect, which does not undergo chemical changes or administers medicine, and it is not a breast implant or a hip, knee, or shoulder joint replacement.

US Medical device classification, as in Europe, depends on the intended use of the device and also on indications for use. Moreover, classification is based on the risk the device poses to the patient and/or the user. There are several factors that may affect the risk including

- the design of the medical device, which should include principles of inherent safety,
- the manufacturing process, which must be well planned, under control, and validated,
- the intended use, which will define the adequate scope of use excluding other places where the medical device is not intended for use,
- the identification of the user, defining its expected experience, education, and training,
- the safety or health of users, implying that the medical device should not compromise the safety of patients.

Most medical devices can be classified by finding the matching description of the device in Title 21 of the Code of Federal Regulations (CFR), Parts 862–892.

The US FDA has established a classification of approximately 1700 generic types of medical devices grouping them in the CFR into several medical specialties referred to as panels:

	Medical Specialty	Regulation Citation (21 CFR)
73	Anesthesiology	Part 868
74	Cardiovascular	Part 870
75	Chemistry	Part 862
76	Dental	Part 872
77	Ear, nose, and throat	Part 874
78	Gastroenterology and urology	Part 876
79	General and plastic surgery	Part 878
80	General hospital	Part 880
81	Hematology	Part 864
82	Immunology	Part 866
83	Microbiology	Part 882
84	Neurology	Part 884
85	Obstetrical and gynecological	Part 886
86	Ophthalmic	Part 888
87	Orthopedic	Part 864
88	Pathology	Part 890
89	Physical medicine	Part 892
90	Radiology	Part 862
91	Toxicology	Part 868

For each of the devices classified by the US FDA, the CFR gives a general description including the intended use, the class to which the device belongs, and information about marketing requirements. Therefore, the panel examines and classifies the device in three different classes of medical devices based on the level of control necessary to assure the safety and effectiveness of the device:

1. **Class I** *(Low Risk)—General Controls:*
   (a) FDC Act lists general references to control the medical devices.
   (b) Some general controls include the following: the device cannot be adulterated or misbranded; the firm must be registered with the US FDA, must maintain records and reports, and must apply good manufacturing practices, etc.
2. **Class II** *(Medium Risk)—Special Controls:*
   (a) For those devices, general controls are not sufficient; therefore, special controls are set.
   (b) Special controls include performance standards; postmarket surveillance; patient registries; guidelines; etc.

3. **Class III** *(High Risk)—Premarket Approval:*

   (a) For those devices, general and special controls are not sufficient; therefore, premarket approval is needed.

   (b) Applications for premarket approval include reports about the safety and effectiveness of the device; a statement of components, properties, and elements of the device; description of the methods, manufacturing controls, packing; references to any relevant standard; sample of the device and components; proposed labeling; certification related to clinical trials; etc.

Thus, the class to which the medical device is assigned determines, among other things, the type of premarketing submission/application required for US FDA clearance to market. However, there are exceptions and exemptions for certain devices.

1.2 NEED FOR MEDICAL DEVICES

Medical devices are indispensable for effective prevention, diagnosis, treatment, and rehabilitation of illness and disease. They help to not only save and prolong life but also improve the quality of life. Therefore, identifying diseases, disabilities, and risk factors is a decisive step to develop new and efficient medical devices. However, besides medical and technological attributes, the development of medical devices is often influenced by other considerations such as markets, costs, and physician preferences.

Nowadays, there are more than 1.5 million different medical devices, including thermometers, surgical drapes, pacemakers, infusion pumps, heart-lung machines, dialysis machines, artificial organs, implants, prostheses, corrective lenses, etc. Currently, orthopedic implants make up the bulk of all devices implanted (~1.5 million per annum worldwide) at a cost of around $10 billion. However, innovation will continuously serve as the fuel for market growth, bringing disruptive products and technologies to market.

New discoveries in biomaterials, technologies, computing, and biology will generate knowledge and growth of new treatments and cures, driving the medical device market to more cost-effective and patient-centered solutions.

Biomaterials are extremely linked to the performance of medical devices, and therefore, to the quality of life of patients. The definition of biomaterials has changed over time [1, 2] while several generations have been developed. However, these generations should be interpreted as the evolution of the requirements and properties of the medical devices. We can group them as follows:

- *Inert Biomaterials:* During the 1960s and 1970s, a first generation of biomaterials was developed for implantation and generation of medical devices. The goal of these inert biomaterials was to replace damaged tissue and provide structural support with a minimal tissue response in the host [3].

- *Bioactive Biomaterials:* In the 1980s and 1990s, a second generation of biomaterials, which was able to elicit a specific biological response at the interface of
  the material, began to develop. The bioactivity was accomplished by using coatings or similar strategies in order to increase and improve implant lifetime by
  optimizing the interface with the host tissue. These bioactive materials allowed
  the creation of more effective and less invasive medical devices [4–6]. Nowadays, this type of biomaterials is still used in many commercial products, for
  example, in dentistry and orthopedics [7].
- *Biodegradable Biomaterials:* Besides the advantages of bioactive materials,
  long-term implants were generally associated with infections, reactions due to
  toxicity or immunological processes, mechanical implant failure due to fatigue,
  etc. As a consequence, a third generation of biomaterials was developed.
  These biomaterials have the capability to degrade and be absorbed offering the
  possibility to overcome the drawbacks of permanent implants [3].
- *Smart Biomaterials:* Progress in biology, proteomics, and bioengineering during the last decade has led to the development of a fourth generation of biomaterials [8–12]. Smart biomaterials are willing to mimic nature's hierarchical
  structures and mechanisms to actively repair and regenerate damaged tissue by
  stimulating specific cellular responses. However, the re-creation of the tissue
  extracellular matrix is complicated and represents one of the challenges of the
  biomaterials field. However, important progress has been made in the design and
  manufacturing of scaffolds for tissue engineering and regenerative medicine.
  Nevertheless, despite their considerable advantages, only few smart biomaterials are being used for clinical applications so far.

While the first generation of biomaterials is still used in a wide range of applications, smart biomaterials will open innovative and new possibilities of treatments and
applications.

Concerning the type of material, the vast majority of medical devices (stents,
orthopedic implants, bone fixators, artificial joints, etc.) are made of metallic materials due to their strength, toughness, and durability. An extensive review on this
topic has been done by Hanawa [13]. Specifically, metals have high strength, high
elasticity, high fracture toughness, and high electrical conductivity when compared
with ceramics and polymers. However, improvements of corrosion resistance and
mechanical durability are needed in order to avoid environmental and health concerns over heavy metals used for medical purposes. In this context, there is a need
to research and improve mechanical and surface properties of metals, because these
features are key to tissue compatibility. Therefore, their physical properties (mechanical, biodegradable, magnetic, etc.) must be improved by redesigning metallic alloys
and biofunctionalizing their surface.

Titanium alloys, such as Ti-6Al-7V, Ti-6Al-7Nb, Ti-6Al-2.5Fe, Ti-13Zr-13Ta,
Ti-6Al-2Nb-1Ta, and Ti-15Zr-4Nb-4Ta, have been widely used for medical and
dental applications. This type of $\alpha + \beta$ titanium alloy shows high corrosion resistance, specific strength, and good tissue compatibility. Because of their Young's
modulus and corrosion resistance, titanium alloys are preferred over stainless steel

and cobalt-chromium alloys for orthopedic and dental applications. However, their low elongation is often associated with fractures. Thus, the development of $\alpha + \beta$ titanium alloys with high elongation and sufficient strength is needed because no optimal titanium alloy is available so far. In addition, Young's modulus of metallic materials is still higher than cortical bone, inducing stress shielding and fracture. To solve these problems, metals with lower Young's modulus are needed. Several β-type titanium alloys with a low Young's modulus, including Ti-12Mo-6Zr-2Fe, T-15Mo, and Ti-15Mo-5Zr-2Al, have been developed for this purpose [14, 15]. On the other hand, ultrahigh-molecular-weight polyethylene and poly(methyl methacrylate) are being used to fill porous titanium alloys in order to obtain materials with reduced Young's modulus [16, 17].

Titanium-nickel alloys have been used to manufacture various medical devices, such as stents, guide wires, and endodontic reamers, due to their specific mechanical properties (shape memory, superelasticity, and damping). Nevertheless, material fractures, pitting, and crevice corrosion have been recently reported [18–21]. In fact, there is a significant problem of toxicity and allergy due to the release of nickel ions. Thus, alloys with better corrosion and fatigue properties must be developed. In fact, there is a huge demand for developing superelastic and shape-memory alloys without using nickel. Some alloys, such as Ti-Sn-Nb, Ti-Mo-Sn, Ti-Nb-O, and Ti-Nb-Al, have been produced with good shape-memory but not enough recovery strain and superelastic deformation stress. On the other hand, nickel-free austenitic stainless steel materials are being developed to obtain materials with better corrosion resistance and strength for medical purposes. Some examples are Fe-(19-23)Cr-(21-24) Mn-(0.5-1.5)Mo-(0.85-1.1)N alloy (BioDur$^{\circledR}$108), Fe-18Cr-18Mn-2Mo-0.9N alloy, and the Fe-(15-18)Cr-(10-12)Mn-(3-6)Mo-0.9N alloy. Co-Cr alloys exhibit excellent corrosion resistance and good wear resistance. However, when using these alloys as orthopedic prosthesis, they produce stress shielding in the adjacent bone. The lack of mechanical stimuli on the bone may lead to the failure and loosening of the implant due to bone resorption. Therefore, osseointegration is another relevant requirement for metallic implantable devices. To solve this problem, there is a need to develop techniques and methodologies to modify their surface in order to give them biofunctionality and improve tissue compatibility. This requirement is currently being fulfilled by using dry and wet processes, which are the most conventional and predominant surface modification techniques [22, 23]. Research in this field is ongoing for techniques that involve the immobilization of biofunctional molecules. However, due to difficulties in ensuring appropriate safety, quality, and durability of those treatments, these are still not used commercially. Further research is needed in order to study the biofunctionalization of metallic materials to use them in innovative technologies such as tissue engineering.

Another important requirement for some medical devices is the ability to be absorbed by the body. This feature is typical of some polymeric materials, but not metallic ones. In fact, there are two metallic materials that should be considered bioabsorbable: iron and magnesium. However, strict control of corrosion rates must be achieved for biodegradable magnesium alloys due to a certain degree of late recoil and neointima formation.

Metallic materials, such as stainless steels, Co-Cr alloys, and titanium alloys, become magnetized when a magnetic field is applied inducing the appearance of artifacts and the disablement of the magnetic resonance imaging (MRI) diagnostic tool. Since this is an important and widely used diagnostic tool, there is a need to develop medical devices made of materials with low magnetic susceptibility. In this sense, materials such as Au-Pt-Nb, Ti-Zr, Zr-Nb, and Zr-Mo alloys are being proposed due to their reduced magnetic susceptibility compared with other material such as Co-Cr-Mo and Ti alloys [24, 25]. However, some of them are difficult to process because of their tensile strength and elongation rates.

Alumina, zirconia, and porous ceramics are commonly employed to develop implantable medical devices such as femoral heads and hip prostheses. Their microstructure depends on the manufacturing system employed and is proportional to the mechanical and biological properties. Ceramic biomaterials show good wear rates, corrosion resistance, biocompatibility, and high strength [7]. Nevertheless, there is a need to increase the quality of ceramic materials to improve their low fracture toughness. On the other hand, porous ceramics (e.g., hydroxyapatite) used to mimic trabecular bone are exposed to mechanical collapse risk, and also their compression strength can be affected by aging. Bioactive ceramics, such as bioactive glasses (BGs), glass-ceramics, and calcium phosphates (CaPs), have been used as bone substitutes for decades due to their similitude with bone mineral structure. However, owing to their low tensile strength, poor mechanical properties, and low fracture toughness, they cannot be used for load-bearing applications. Further studies are needed to improve the mechanical features of these kinds of ceramic materials.

On the other hand, the use of polymers in surgery, prosthetics, pharmacology, and drug delivery is essential. Many polymeric compounds are considered biomaterials and used in many applications: silicones (tubes, plastic surgery), polyurethanes (catheters, cardiac pumps), polytetrafluoroethylene (orthopedics), nylon-type polyamides (sutures), polymeric compounds based on methyl methacrylate (cements, odontology, prostheses), etc. However, there is a need to improve their biostability and performance in terms of clinical applications because the release of wear debris is often present in those materials leading to undesirable effects.

Many medical devices are implanted in the body and a second surgical procedure is often required to remove the remnants of a previous implant. In this regard, the use of biodegradable polymers has been the key [26]. Moreover, their flexibility, durability, and biocompatibility have made them very important to develop safety devices such as orthopedic implants, sutures, drug eluting stents, and scaffolds. Moreover, biodegradable polymers can reduce the stress shielding effect, avoid removal of implants, and enable postoperative diagnostic imaging. Biodegradable polymers can be classified as synthetic and natural. Synthetic polymers are able to be hydrolyzed by human tissues. Some examples are polylactic acid (PLA), polyglycolic acid (PGA), poly-ε-caprolactone (PCL), polyethylene glycol (PEG), etc. [27]. Natural biodegradable polymers are proteins or polysaccharides (chitosan, agarose, collagen, alginate, etc.), which undergo enzymatic degradation. The mechanisms of polymeric degradations are bulk erosion and surface erosion. The understanding of these mechanisms is crucial to design and develop safe and efficient smart devices, drug delivery devices, scaffolds, or bioactive products. From a mechanical point of view, these types of polymers can be reinforced using oriented fibers or fibrils of the same material. This strategy provides these materials with fair mechanical properties.

Another type of frequently used polymeric material for medical device applications is silicone. Silicones have unique properties, such as biocompatibility, biodurability, chemical and thermal stability, hydrophobicity, and low surface tension, which allow them to be extensively used in the medical field. Silicones have been used to manufacture orthopedic implants, catheters, drains, shunts, medical machines, valves, esthetic implants, stents, to name just a few. However, there are some concerns about the biocompatibility and biodurability of silicones that have recently been under discussion. Some silicone materials have not passed the biocompatibility tests, and their purity is not suitable for medical purposes. On the other hand, notwithstanding silicones' chemical stability, certain factors may affect its durability and long-term performance.

Based on clinical experience, there are several biological, mechanical, chemical, and physical requirements for biomaterials that should be targeted to develop more efficient and adequate medical devices including foreign body reaction (due to wear fibrils), stress shielding, biocompatibility, bioactivity, osteoinduction, etc. A description of the major requirements is listed here:

- *Safety:* it is the most important requirement for medical devices. They must be safe and not show any toxicity. Therefore, corrosion-resistant materials should be used.
- *Durability:* there is a need to improve the durability of materials and wear resistance in order to increase product life and reduce medical interventions due to replacement or fatigue problems.
- *Mechanical Compatibility:* this is a key characteristic for multiple purposes such as to avoid stress shielding.
- *Biodegradability:* in order to increase biocompatibility, reduce immune reactions, and avoid retrieval, there is a need to develop biodegradable materials.
- *Biofunction:* to improve the performance of several medical devices, there is a need to promote bone formation (e.g., fixation of devices in bone), to promote adhesion of soft tissue (e.g., fixation of soft tissue), to prevent thrombus (e.g., inhibition of platelet adhesion), to avoid infections (e.g., inhibition of biofilm formation), to reduce magnetic susceptibility (e.g., avoid artifacts in MRI), etc.

Finally, a major concern in the medical device field is infection. Bacteria often colonize the surface of medical devices developing a biofilm that compromises not only the functionality and performance of the device but also the patient's health. For these cases, removal of the infected device is frequently the only option. Therefore, a solution for septic failures must be considered as a real need due to the high morbidity and enormous costs associated. Currently, there is no specific approach that can ensure the development of medical devices exempt from possible infections. Even sterile procedures, antibiotic prophylaxis, and appropriate aseptic management of the devices do not guarantee obtaining a resistant device. However, along with preventive measures, the use of biomaterials with a certain degree of bacterial resistance is being applied. There are many materials, such as noble metals, that have bactericidal properties. With the development of new medical devices, an increased number of anti-infective biomaterials has arisen. Biomaterials have been formulated to release antibacterial, antifungal, antiviral, antiprotozoal, and anthelmintic drugs. And even to treat other types of pathogens. The vast majority of anti-infective biomaterials are the ones with antibacterial properties [28]. Some of the most broadly used medical devices, such as contact lenses and catheters, are extremely exposed to those infections. However, the potential risk of infection is based on the degree of invasiveness of each medical device. Thus, the strategy to determine the most suitable anti-infective device should take it into account. Whether we have external medical devices, partially internal medical devices, or a totally internal medical devices will determine the type of possible infections and the characteristics of the process. Future strategies should focus on designing and developing biomaterials with specific and appropriate anti-infective properties for each application.

The complexity of new biomaterials demands to develop specific and efficient manufacturing technologies. Current needs in biomaterials and technologies should pursue the goal of enhancing tissue regeneration rather than replacement. To address this need, several strategies are being explored including surface modification, development of drug delivery systems, generation of advanced three-dimensional scaffolding geometries for tissue engineering, etc. Moreover, some researchers are willing to develop smart materials with intrinsic ability to enhance and promote the capacity of the damaged tissue to self-repair and regenerate by stimulating cellular migration, attachment, and proliferation, as well as vascularization and nutrient supply [29–33]. These capabilities can be imparted to biomaterials by using technologies that allow modifying or regulating some surface characteristics (e.g., roughness), morphological properties (e.g., porosity), degradation mechanisms, and mechanical features. Nevertheless, besides the progress made in the field, a major barrier to promote this innovation is the low rate of commercially available smart medical devices.

1.3   TECHNOLOGY CONTRIBUTION TO MEDICAL DEVICES

Design and manufacturing of medical devices is essential to improve patients' quality of life and treatment effectiveness while reducing health-care costs. Owing to the nature of these devices, the requirements to develop medical devices include biocompatibility, reliability, corrosion resistance, controllability, and customization among others. Currently, conventional technologies are being used to obtain marketable products; however, advanced manufacturing technologies are required to achieve innovative medical devices with desired results.

To meet the challenges of new medical devices, the term Biomanufacturing has been addressed. Biomanufacturing was defined as the application of design and manufacturing technologies to reduce the cost while advancing the safety, quality, efficiency, and speed of health-care service and biomedical science [34]. The term includes several fields of study such as design, mechatronics, fabrication, and assembly.

The technology contribution to medical devices is based on the combination of a good understanding and availability of materials and fabrication processes. Some requirements should be first addressed, such as, biocompatibility and corrosion/fatigue resistance. For this reason, the election of an appropriate material and manufacturing technology is very important.

A new set of manufacturing technologies emerged in the past decades to address market requirements such as the need to develop customized low-cost medical devices. These new technologies are usually referred to as rapid prototyping [35]. Rapid prototyping (RP), developed around the mid-1980s, enables the production of useful prototypes to test the fit, form, and function of medical devices prior to their release to the market. Nowadays, these technologies have evolved to rapid manufacturing (RM) and are able to produce directly functional parts and products. In addition, RP is also useful to create models to plan and prepare surgical approaches or to train medical students.

The medical device manufacturing technologies can be classified into subtractive, net-shape, and additive processes. A brief description of some of these technologies is given in the following sections. More detailed information is presented in the following chapters.

1.3.1 Subtractive Technologies

The progress in medical devices based on new technologies is often transferred from other industries to the medical field. Within subtractive technologies, we find mechanical (turning, milling, drilling, and grinding), electric (EDM), electronic beam, and chemical processes. All of them have been used to manufacture a broad variety of medical devices.

Machining technologies have been used to manufacture some medical devices. In particular, laser micromachining has been widely used to produce vascular stents [36]. A stent is a tubular wire mesh that is deployed in a diseased coronary artery to provide a smooth blood circulation [37]. Recently, the emergence of fiber laser technologies has enabled their increasing applications in medical device micromachining. In laser machining, device fabrication is performed by incremental removal of material using a laser. However, ablation of material cannot take place at locations where the laser path is obstructed. Therefore, this constraint limits the geometries that can be fabricated by laser machining, such as structures with overhangs or interior geometries [38]. Other applications of laser machining are the production of scaffolds for tissue engineering and channels for microfluidic devices [39, 40].

1.3.2 Net-Shape Technologies

Forming and molding are used for large-scale production. In the case of medical device production, for example, pacemakers and electronic connections are made using forming processes, while hearing aid devices are molded. For small batches or single products, a new process has been recently used: incremental sheet forming (ISF) [41, 42]. Due to the natural difference between individuals and treatments, incremental forming technologies can also be applied to develop new medical devices. This process will allow high customization with affordable costs. Thus, research has been focused on increasing the geometric accuracy [43–45], geometric flexibility [46, 47], and exploring new applications [48–50].

1.3.3 Additive Technologies

Over the past decades, additive technologies have been used to produce medical devices from CAD models based on computed tomography (CT), MRI, and other medical imaging techniques [51–53]. In fact, the number of applications for these technologies in the medical area is increasing. The basic concept of additive manufacture is the fabrication of a 3D model by adding consecutive layers. The main advantage of this technique is the capacity to rapidly produce very complex 3D models and the ability to use various raw materials. Using additive technologies, it is possible to obtain customized devices with specific geometry, size, and function for a given patient and a given medical context. In addition, surgeons can use additive processes to fabricate exact anatomical replicas for surgical planning or training. Additive processes can be divided in electrochemical processes and physical processes.

1.3.3.1 Electrospinning Electrospinning uses an electrostatically driven jet of polymer solution to produce micro- or nano-fibers. These fibers can be used to produce nanostructures for tissue engineering [54–58].

1.3.3.2 Stereolithography Stereolithography (SLA) is an additive process used to produce 3D objects through curing a photoreactive resin with a power source [59–61]. Excitation of photoreactive molecules, formation of reactive species, generation of free radicals, and polymerization of the resin occurs in the region of laser-resin interaction. Two distinct methods are usually employed in this process: mask-based method and UV beam method. In the former, an image is transferred to a liquid polymer by irradiation through a patterned mask, while in the latter, a UV beam is focused to selectively solidify the liquid resin. SLA is used to produce surgical guides for the placement of dental implants, temporary crowns and bridges, and resin models for lost wax casting, among others [62]. SLA of biodegradable polymers has also been used to produce tissue engineering scaffolds [63, 64]. Several clinically implanted prostheses, including auricular, maxillofacial, and cranial prostheses, have been obtained using this technology [65–67], as well as hydroxyapatite (HA) ceramic scaffolds for orbital floor prosthesis [68].

1.3.3.3 Extrusion-Based Techniques The extrusion-based technique is commercially known as fused deposition modeling (FDM). This process was developed by Crump and marketed by Stratasys Corporation [69]. In this process, a thin filament of plastic material is melted and extruded through an extrusion head, which deposits the filament layer by layer to produce a final part. This FDM process is limited to available thermoplastic materials that may be formed into filaments, heated, and deposited. However, they have been used to produce parts from medical-grade materials. Nowadays, the use of FDM to produce permanent implants has been limited to the fabrication of models that can serve as templates for the fabrication of custom implants [70]. On the other hand, extrusion-based processes (bioextrusion) are widely used to produce scaffolds for tissue engineering.

Another extrusion-based technique has appeared recently. The multi-phase jet solidification (MJS) is based on the extrusion of metal or ceramic slurries using the metal injection molding technique. The main difference between this technique and the FDM lies in the raw material and the feeding system [71].

1.3.3.4 Inkjet Printing Inkjet printing deposits tiny droplets of material onto the required locations to form layer by layer a 3D object. The material is a mixture of the raw material with a binder, which is removed later. Inkjet printing comprises two configurations: the bonder method and the direct build-up method. The most common application of inkjet printing is the production of prototypes for form and fit testing. However, other medical devices are being produced using this kind of technology, including MEMS, drug eluting devices, scaffolds, and artificial tissues and organs.

1.3.3.5 Laser Sintering and Melting Selective laser sintering (SLS) and selective laser melting (SLM) are additive manufacturing processes that use high-energy light sources to consolidate powder material [72–75]. Both technologies have been used to produce permanent and temporary implants [76].

1.3.3.6 Electron Beam Melting Electron beam melting (EBM) is an additive manufacturing process that uses an electron beam to scan a layer of metal powder on a substrate, forming a melt pool [77, 78]. The system consists of the electron beam gun compartment and the specimen-fabrication compartment both kept in a high vacuum. EBM has been used to produce titanium root-form implants (Ti-6Al-4V ELI) [79], femur-hip implants [80], dental implants [77], and knee replacement implants [81].

Additive manufacturing (AM) technologies, as an evolution of rapid prototyping methods, can produce direct products satisfying mechanical, physical, chemical, and biological requirements.

In the medical area, AM are playing a key role due to its capability to reproduce complex shapes and geometries impossible to be produced using other manufacturing processes. There are many research works showing how additive technologies can produce difficult shapes, adjust process parameters, or check mechanical capabilities [82]. In general, research is focused on additive manufacturing technologies performance using different technologies and different materials. Chimento et al. [83] evaluated the performance of 3D printed materials for use as rapid tooling (RT) moulds in low-volume thermoforming processes such as in manufacturing custom prosthetics and orthotics.

1.4   CHALLENGES IN THE MEDICAL DEVICE INDUSTRY

The aging of the population, the need to have a good quality of life, and altogether the scientific and technological progress represent an immense potential for the medical device market. This situation creates an ideal environment for the emergence of new challenges and opportunities. In this section, some challenges of the medical devices industry are explained.

- Innovation in the medical device market implies improving the quality of treatment, considering the quality of established treatment options (e.g., reduce morbidity and mortality, reduce pain, increase implant survival, etc.), while improving the costs of the treatment, considering the direct and indirect costs of the established treatments. Innovation seeks a beneficial effect in both pillars, increasing quality while simultaneously decreasing costs. However, many recently developed smart medical devices increase costs, mainly due to the enormous associated technological knowledge. Consequently, one major challenge is to develop appropriate industrial technologies and processes capable of producing reproducible, safe, effective, and economically acceptable medical devices.
- Due to the globalization of the markets, health-care providers are exploring new opportunities in developing nations in order to increase efficiencies and reduce costs. According to the United Nations, there are approximately 5 billion people living in low-resource developing countries, while only 1.2 billion live in high-resource developed nations. This means that the potential market in developing countries is enormous, generating new opportunities for lower production costs. Therefore, a challenge of the medical device industry is to explore and exploit this market by exporting medical devices or directly establishing manufacturing facilities in low-income countries. However, companies must pay attention to maintaining workers' rights and ethical issues.
- Important challenges are associated with medical devices' policies and regulations. New strategies need to be developed to handle the generation of new medical devices such as regenerative medical devices (scaffolds) and drug-eluting medical devices among others. Nevertheless, it is essential for the collaboration of all stakeholders (medical device industry, governments, academia, NGOs, professionals, etc.) to share knowledge and expertise in order to determine specific and required actions to increase the quality, effectiveness, and coverage of health-care policies and regulations.
- Personalized or customized treatments have proven to be more efficient, effective, and safe in many cases. Although the generation of customized products is a hot topic in the biomedical research area, its application in the market is very limited due to regulatory and practical reasons. Nowadays, the design and manufacturing of medical devices are subject to national and international policies and regulations that require clinical trials to validate their efficacy and safety. Therefore, any deviation concerning the design, materials, or manufacturing

procedures require extensive studies and tests to be re-approved. This is a clear barrier against the generation of customized medical devices and represents a challenge to overcome in order to produce more suitable and efficient products and treatments.

- A current focus of research is the development of smart medical devices. Smart medical devices are made of bioinspired materials able to reproduce the function of the tissue or body part that they are replacing or assisting. They have complex and multifunctional structures that can integrate multiple chemical inputs and physical stimuli that determine their behavior. Such devices could target desired anatomical regions or specific cellular populations, making it a very effective treatment with minimal side effects. The study and generation of these devices and biomaterials is both a challenge and a need. The understanding of biological processes, the creation of new biomaterials, and the development of innovative technologies are a requirement to create devices for diagnostic or therapeutic applications.

- Allografts, autografts, and xenografts are currently being used to replace damaged tissue. Nevertheless, there are several limitations associated with them, such as scarcity, rejection, disease transfer, harvesting costs, and postoperative morbidity to name a few [84–86]. In those situations, tissue engineering or regenerative medicine have emerged as an effective and adequate alternative to overcome these problems. Tissue engineering is based on the use of scaffolds combined with natural components (e.g., cells, growth factors, molecules, etc.) and signaling pathways to repair and regenerate organs and tissues [87]. Although there are many research areas exploring and studying tissue engineering, its implementation as a routine treatment is still controversial. A significant challenge to solve is the angiogenesis, which is crucial to stimulate tissue regeneration after implantation. In addition, the manipulation of these structures, such as the scaffolds with cells, represents an additional complication for their widespread use in hospitals.

REFERENCES

[1] Black J, Hastings G, editors. *Handbook of Biomaterial Properties*. London: Chapman & Hall,
Thomson Science; 1998.

[2] Williams DF. On the nature of biomaterials. Biomaterials 2009;30:5897–5909. Priority medical devices for specific purposes.

[3] Hench LL. Biomaterials. Science 1980;208:826–831.

[4] Hench LL. Biomaterials: a forecast for the future. Biomaterials 1998;19:1419–1423.

[5] Hench LL, Wilson J. Surface-active biomaterials. Science 1984;226:630–636.

[6] Burns JW. Biology takes centre stage. Nat Mater 2009;8:441–443.

[7] Navarro M, Michiardi A, Castano O, Planell JA. Biomaterials in orthopaedics. J R Soc Interface 2008;5:1137–1158.

[8] Anderson DG, Burdick JA, Langer R. Materials science. Smart biomaterials. Science 2004;305:1923–1924.

[9] Boyan BD, Schwartz Z. Regenerative medicine: are calcium phosphate ceramics "smart" biomaterials. Nat Rev Rheumatol 2011;7:8–9.

[10] Furth ME, Atala A, Van Dyke ME. Smart biomaterials design for tissue engineering and regenerative medicine. Biomaterials 2007;28:5068–5073.

[11] Mieszawska AJ, Kaplan DL. Smart biomaterials — regulating cell behaviour through signaling molecules. BMC Biol 2010;8:59.

[12] Yuan H, Fernandes H, Habibovic P, de Boer J, Barradas AM, Walsh WR, van Blitterswijk CA, De Bruijn JD. "Smart" biomaterials and osteoinductivity. Nat Rev Rheumatol 2011;7(4):1.

[13] Hanawa T. Research and development of metals for medical devices based on clinical needs. Sci Technol Adv Mater 2012;13:064102.

[14] Wang K, Gustavson L, Dumbleton J. *Titanium '92: Science and Technology*. Warrenda: TMS; 1993. p 2697.

[15] Zardiackas LD, Mitchell DW, Disegi JA. *Medical Applications of Titanium and its Alloys*. West Conshohoken, PA: ASTM; 1996. p 60.

[16] Nomura N, Baba Y, Kawamura A, Fujinuma S, Chiba A, Masahashi N, Hanada S. Mechanical properties of porous titanium compacts reinforced by UHMWPE. Mater Sci Forum 2007;539–543:1033.

[17] Nakai M, Niinomi M, Akahori T, Yamanoi H, Itsuno S, Haraguchi N, Itoh Y, Ogasawara T, Onishi T, Shindo T. Effect of Silane Coupling Treatment on Mechanical Properties of Porous Pure Titanium Filled with PMMA for Biomedical Applications. J Jpn Inst Met 2008;72:839–845.

[18] Scheinert D, Scheinert S, Sax J, Piorkowski C, Braunlich S, Ulrich M, Biamino G, Schmidt A. Prevalence and clinical impact of stent fractures after femoropopliteal stenting. J Am Coll Cardiol 2005;45:312–315.

[19] Kang WY, Kim W, Kim HG, Kim W. Drug-eluting stent fracture occurred within 2 days after stent implantation. Int J Cardiol 2007;120:273–275.

[20] Robertson SW, Ritchie RO. In vitro fatigue-crack growth and fracture toughness behavior of thin-walled superelastic Nitinol tube for endovascular stents: A basis for defining the effect of crack-like defects. J Biomed Mater Res Part B 2008;28(4):700–709.

[21] Gall K, Tyber J, Wilkesanders G, Robertson SW, Ritchie RO, Maier HJ. Effect of microstructure on the fatigue of hot-rolled and cold-drawn NiTi shape memory alloys. Mater Sci Eng, A 2008;486:389.

[22] Hanawa T. An overview of biofunctionalization of metals in Japan J R Soc Interface 2009;6:S361.

[23] Hanawa T. Biological reactions on titanium surface electrodeposited biofunctional molecules. In: Sasano T, Suzuki O, editors. 3rd International Symposium for Interface Oral Health Science; Jan 15–16, 2009; Sendai, Japan; Interface Oral Health Science 2009; 2010. p 83–89.

[24] Nomura N, Tanaka Y, Suyalatu KR, Doi H, Tsutsumi Y, Hanawa T. Effects of Phase Constitution of Zr-Nb Alloys on Their Magnetic Susceptibilities. Mater Trans 2009;50:2466.

[25] Suyalatu NN, Oya K, Tanaka Y, Kondo R, Doi H, Tsutsumi Y, Hanawa T. Microstructure and magnetic susceptibility of as-cast Zr-Mo alloysOriginal Research Article. Acta Biomater 2010;6:1033–1038.

[26] Brannon-Peppas L, Vert M. In: Wise DL, Klibanov A, Mikos A, Brannon-Peppas L, Peppas NA, Trantalo DJ, Wnek GE, Yaszemski MJ, editors. Handbook of Pharmaceutical Controlled Release Technology. New York: Marcel Dekker; 2000. p 99.

[27] Vroman I, Tighzert L. Biodegradable polymers. Materials 2009;2:307–344.

[28] Campoccia D, Montanaro L, Arciola CR. A review of the clinical implications of anti-infective biomaterials and infection-resistant surfaces. Biomaterials 2013;34:8018–8029.

[29] Place ES, Evans ND, Stevens MM. Complexity in biomaterials for tissue engineering. Nat Mater 2009;8:457–470.

[30] Jakob F, Ebert R, Rudert M, Noth U, Walles H, Docheva D, Schieker M, Meinel L, Groll J. In situ guided tissue regeneration in musculoskeletal diseases and aging: implementing pathology into tailored tissue engineering strategies. Cell Tissue Res 2011;347:725–735.

[31] Hutmacher DW. Scaffolds in tissue engineering bone and cartilage. Biomaterials 2000;21:2529–2543.

[32] Hutmacher DW, Schantz JT, Lam CX, Tan KC, Lim TC. State of the art and future directions of scaffold-based bone engineering from a biomaterials perspective. J Tissue Eng Regener Med 2007;1:245–260.

[33] Dormer NH, Berkland CJ, Detamore MS. Emerging techniques in stratified designs and continuous gradients for tissue engineering of interfaces. Ann Biomed Eng 2010;38:2121–2141.

[34] Mitsuishi M, Cao J, Bartolo P, Friedrich D, Shih AJ, Rajurkar K, Sugita N, Harada K. Biomanufacturing. CIRP Ann 2013;62:585–606.

[35] Lantada AD, Morgado PL. Rapid prototyping for biomedical engineering: current capabilities and challenges. Annu Rev Biomed Eng 2012;14:73–96.

[36] Stoeckel D, Bonsignore C, Duda S. A survey of stent designs. Minim Invasive Ther Allied Technol 2002;11(4):137–147.

[37] Whittaker DR, Fillinger MF. The engineering of endovascular stent technology. Vasc Endovascular Surg 2006;40(2):85–94.

[38] Gittard SD, Narayan RJ. Laser direct writing of micro- and nano-scale medical devices. Expert Rev Med Devices 2010;7(3):343–356.

[39] Applegate RW, Schafer DN, Amir W, et al. Optically integrated microfluidic systems for cellular characterization and manipulation. J Opt A: Pure Appl. Opt. 2007;9(8):S122–S128.

[40] Vazquez RM, Osellame R, Cretich M, et al. Optical sensing in microfluidic lab-on-a chip by femtosecond-laser-written waveguides. Anal Bioanal Chem 2009;393(4):1209–1216.

[41] Hirt G, Junk S, Witulsky N. *Incremental Sheet Forming: Quality Evaluation and Process Simulation*, In: Proceeding of the 7th ICTP Conference; Yokohama; 2002. p 925–930.

[42] Park JJ, Kim YH. Fundamental studies on the incremental sheet metal forming technique. J Mater Process Technol 2003;140:447–453.

[43] Araghi BT, Manco GL, Bambach M, Hirt G. Investigation into a new hybrid forming process: incremental sheet forming combined with stretch forming. CIRP Ann 2009;58(1):225–228.

[44] Duflou JR, Callebaut B, Verbert J, De Baerdemaeker H. Laser assisted incremental forming: formability and accuracy improvement. CIRP Ann 2007;56(1):273–276.

[45] Malhotra R, Cao J, Beltran M, Xu D, Magargee J, Kiridena V, Xia ZC. Accumulative-DSIF strategy for enhancing process capabilities in incremental forming. CIRP Ann 2012;61(1):251–254.

[46] Durante M, Formisano A, Langella A, Capece Minutolo FM. The influence of tool rotation on an incremental forming process. J Mater Process Technol 2009;209(9):4621–4626.

[47] Verbert J, Belkassem B, Henrard C, Habraken AM, Gu J, Sol H, Lauwers B, Duflou JR. Multi-step toolpath approach to overcome forming limitations in single point incremental forming. Int J Mater Form 2008;1(S1):1203–1206.

[48] Ambrogio G, Filice L, Manco GL. Warm incremental forming of magnesium alloy AZ31. CIRP Ann 2008;57(1):257–260.

[49] Fan G, Sun F, Meng X, Gao L, Tong G. Electric hot incremental forming of Ti–6Al–4V titanium sheet. Int J Adv Manuf Technol 2009;49(9–12):941–947.

[50] Oleksik V, Pascu A, Deac C, Fleaca R, Roman M, Bologa O, Barlat F, Moon YH, Lee MG. The influence of geometrical parameters on the incremental forming process for knee implants analyzed by numerical simulation. AIP Conf Proc 2010;1252(1):1208–1215.

[51] Solar P, Ulm C, Lill W, et al. Precision of three-dimensional CT-assisted model production in the maxillofacial area. Eur Radiol 1992;2(5):473–477.

[52] Ono I, Gunji H, Kaneko F, Numazawa S, Kodama N, Yoza S. Treatment of extensive cranial bone defects using computer-designed hydroxyapatite ceramics and periosteal flaps. Plast Reconstr Surg 1993;92(5):819–830.

[53] Eufinger H, Wehmoller M, Machtens E, Heuser L, Harders A, Kruse D. Reconstruction of craniofacial bone defects with individual alloplastic implants based on CAD/CAM-manipulated CT-data. J Craniomaxillofac Surg 1995;23(3):175–181.

[54] Agrawal S, Wendorff JH, Greine A. Use of electrospinning technique for biomedical applications. Polymer 2008;49:5603–5621.

[55] Barnes CP, Sell SA, Boland ED, Simpson DG, Bowlin GL. Nanofiber technology: designing the next generation of tissue engineering scaffolds. Adv Drug Delivery Rev 2007;59:1413–1433.

[56] Ekaputra AK, Zhou Y, Cool SM, Hutmacher DW. Composite electrospun scaffolds for engineering tubular bone grafts. Tissue Eng Part A 2009;15:3779–3788.

[57] Moghe AK, Gupta BS. Co-axial electrospinning for nanofiber structures: preparation and applications. Polym Rev 2008;48:353–377.

[58] Mitchell GR, Ahn KH, Davis FJ. The potential of electrospinning in rapid manufacturing. Virtual and Physical Prototyping 2011;6:63–77.

[59] Bartolo PJ. Photo-curing modeling: direct irradiation. Int J Adv Manuf Technol 2007;32:480–491.

[60] Bartolo PJ, Gaspar J. Metal filled resin for stereolithography metal part. CIRP Ann 2008;57(1):235–238.

[61] Bartolo PJ, Lenz E. Computer simulation of stereolithographic curing reactions: phenomenological versus mechanistic approaches. CIRP Ann 2006;55(1):221–225.

[62] Noort RV. The future of dental devices in digital. Dent Mater 2012;28:3–12.

[63] Lee JW, Nguyen TA, Kang KS, Seol Y, Cho D. Development of a growth factor-embedded scaffold with controllable pore size and distribution using microstereolithography. Tissue Eng A 2008;14(5):835.

[64] Lee JW, Lan PX, Kim B, Lim G, Cho D. 3D scaffold fabrication with PPF/DEF using micro-stereolithography. Microelectron Eng 2007;84(5–8):1702–1705.

[65] Subburaj K, Nair C, Rajesh S, Meshram SM, Ravi B. Rapid development of auricular prosthesis using CAD and rapid prototyping technologies. Int J Oral Maxillofac Surg 2007;36(10):938–943.

[66] Singare S, Lian Q, Wang WP, et al. Rapid prototyping assisted surgery planning and custom implant design. Rapid Prototyping J 2009;15(1):19–23.

[67] Wurm G, Tomancok B, Holl K, Trenkler J. Prospective study on cranioplasty with individual carbon fiber reinforced polymer (CFRP) implants produced by means of stereolithography. Surg Neurol 2004;62(6):510–521.

[68] Levy RA, Chu TGM, Holloran JW, Feinberg SE, Hollister S. CT-generated porous hydroxyapatite orbital floor prosthesis as a prototype bioimplant. Am J Neuroradiology 1997;18:1522–1525.

[69] Crump SS. Apparatus Method for Creating Three-dimensional Objects. US Patent 5121329. 1989.

[70] Bartolo P, Kruth JP, Silva J, Levy G, Malshe A, Rajurkar K, Mitsuishi M, Ciurana J, Leu M. Biomedical production of implants by additive electro-chemical and physical processes. CIRP Ann 2012;61:635–655.

[71] Greulich M, Greul M, Pintat T. Fast, functional prototypes via multiphase jet solidification. Rapid Prototyping J 1995;1(1):20–25.

[72] Averyanova M, Bertrand P, Verquin B. Manufacture of Co–Cr dental crowns and bridges by selective laser melting technology. Virtual Phys Prototyping 2011;6:179–185.

[73] Goodridge RD, Tuck CJ, Hague RJM. Laser sintering of polyamides and other polymers. Prog Mater Sci 2012;57:229–267.

[74] Klocke F, Wagner C. Coalescence of two metallic particles as base mechanism of selective laser sintering. CIRP Ann 2003;52(1):177–184.

[75] Yadroitsev I, Smurov I. Surface morphology in selective laser melting of metal powders. Phys Procedia 2011;12:264–270.

[76] Vandenbroucke B, Kruth J-P. Direct Digital Manufacturing of Complex Dental Prostheses. In: Bartolo P, Bidanda B, editors. *Bio-materials and Prototyping Applications in Medicine*. Springer; 2008.

[77] Koike M, Martinez K, Guo L, Chahine G, Kovacevic R, Okabe T. Evaluation of titanium alloy fabricated using electron beam melting system for dental applications. J Mater Process Technol 2011;211:1400–1408.

[78] Mazzoli A, Germani M, Raffaeli R. Direct fabrication through electron beam melting technology of custom cranial implants designed in a PHANToM-based haptic environment. Mater Des 2009;30:3186–3192.

[79] Chahine G, Koike M, Okabe T, Smith P, Kovacevic R. The design and production of Ti–6Al–4V ELI customized dental implants. JOM 2008;60:50–55.

[80] Harrysson OL, Cansizoglu O, Marcellin-Little DJ, Cormier DR, West IIHA. Direct metal fabrication of titanium implants with tailored materials and mechanical properties using electron beam melting technology. Mater Sci Eng, C 2008;28:366–373.

[81] Murr LE, Amato KN, Li SJ, Tian YX, Cheng XY, Gaytan SM, Martinez E, Shindo PW, Medina F, Wicker RB. Microstructure and mechanical properties of open-cellular biomaterials prototypes for total knee replacement implants fabricated by electron beam melting. J Mech Behav Biomed Mater 2011;4:1396–1411.

[82] Delgado J, Ciurana J, Rodríguez CA. Influence of process parameters on part quality and mechanical properties for DMLS and SLM with iron-based materials. Int J Adv Manuf Technol 2012;60:601–610.

[83] Chimento J, Highsmith MJ, Crane N. 3D printed tooling for thermoforming of medical devices. Rapid Prototyping J 2011;17:387–392.

[84] Fernyhough JC, Schimandle JJ, Weigel MC, Edwards CC, Levine AM. Chronic donor site pain complicating bone graft harvest from the posterior iliac crest for spinal fusion. Spine 1992;17:1474–1480.

[85] Banwart JC, Asher MA, Hassanein RS. Iliac crest bone graft harvest donor site morbidity. A statistical evaluation. Spine 1995;20:1055–1060. (doi: 10.1097/00007632-199505000-00012).

[86] Goulet JA, Senunas LE, DeSilva GL, Greengield MLVH. Autogenous iliac crest bone graft. Complications and functional assessment. Clin Orthop 1997;339:76–81. (doi: 10.1097/00003086-199706000-00011).

[87] Hardouin P, Anselme K, Flautre B, Bianchi F, Bascoulenguet G, Bouxin B. Tissue engineering and skeletal diseases. Joint Bone Spine 2000;67:419–424.

2

DESIGN ISSUES IN MEDICAL DEVICES

Inés Ferrer and Jordi Grabalosa

Department of Mechanical Engineering and Industrial Construction, University of Girona, Girona, Catalonia, Spain

Alex Elias-Zuñiga and Ciro Angel Rodriguez

Department of Mechanical Engineering, Instituto Tecnológico y de Estudios Superiores de Monterrey, Campus Monterrey, Monterrey, Nuevo León, Mexico

2.1 MEDICAL DEVICE DEVELOPMENT (MDD)

Product development means to create products with new or different features to satisfy different customer needs or to improve the satisfaction of existing ones [1]. During this process, there is a transformation of information from a need, requirements, and restrictions to a physical structure that is able to meet these demands [2]. From a general perspective, product development is an interaction between "what" is to be achieved and "how" it is to be achieved [3].

Focusing on medical devices, product development is more difficult than the development of conventional products. Basically, in the health sector, there is a huge variety of regulations, standards, procedure, and controls that these types of products have to follow and overcome. Before being introduced into the market, designers have to guarantee complete product safety—both the satisfaction of all the requirements related to the product and the safety and effectiveness of the manufacturing process and quality control procedures [1]. The regulations are different depending on the country and all medical devices sold in a country must conform to the local

Biomedical Devices: Design, Prototyping, and Manufacturing, First Edition.
Edited by Tuğrul Özel, Paolo Jorge Bártolo, Elisabetta Ceretti, Joaquim De Ciurana Gay,
Ciro Angel Rodriguez, and Jorge Vicente Lopes Da Silva.
© 2017 John Wiley & Sons, Inc. Published 2017 by John Wiley & Sons, Inc.

regulations. These regulations protect consumer's health and safety, leading to the development process becoming more rigorous.

The medical product development sector has many opportunities, although it faces a number of challenges [4]. Growing competition in either drugs or devices, in several issues of the process development, is a prominent challenge. Medical device companies have to be flexible and innovative to guarantee the success in the market, assure customer satisfaction, and fulfill all the requirements of this highly regulated and controlled industry [4]. The reduced product cost, the faster product time development, improved features, and reliable high-quality products to achieve higher customer satisfaction are taking relevance in this sector. The trend on increasing the product complexity and the miniaturization of several medical devices leads to the design of small parts with complex geometry and tight tolerances that strongly affects the manufacturing technologies and automation techniques. Moreover, the huge quantity of functions that the products has to satisfy implies the use of more advanced components and leads to the development of new biocompatible materials to achieve high levels of performance [4].

The investments required to develop medical devices are quite high, and sometimes the device's life cycle and batch size are not enough to recover them [5]. Moreover, it is well known that many products fail to fulfill the specifications during their first tests and consequently more investments may be needed.

In the medical device industry, there are a huge variety of users who use the devices differently and with dissimilar expectations, and have distinctive perspectives of them. Often the intermediate or end user is not directly responsible for the payment of the device or treatment. It is important to differentiate between the terms *customer* and *user* [1]. Customer can be defined as a person or company who purchases goods or services independently of benefiting from them. They may be governments, companies, or engineers. The user is a person who uses a medical device for the treatment and/or care of him-/herself or someone else. They may be health-care professionals, patients, caregivers, or people with special needs, among others [1]. Both users and customers have to be involved in the development of medical devices to guarantee the success of the final product.

2.1.1 Biomedical Product Life Cycle

The life cycle of the biomedical products is extended compared with conventional products, although more issues have to be considered for developing them. Figure 2.1 shows a cyclic model of the different phases of the medical product life cycle CDRH and US FDA 2011 [6]. Similar to conventional products, the phases go from the creation of the concept to the retreat, and involve disciplines such as design, manufacturing process planning, quality control, production management, maintenance, after-sales service, and postmarketing surveillance. However, for biomedical products, it is necessary to develop the prototype, preclinical, and clinical analyses in order to test, improve, retest, validate, optimize, and finalize before beginning the manufacturing process. The more complicated ones due to the strong verification

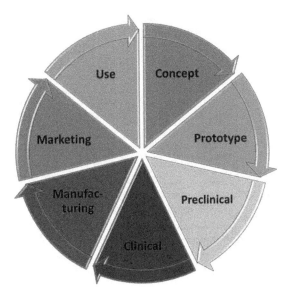

Figure 2.1 Total medical device life cycle (according to Center for Devices and Radiological Health (CDRH), US FDA).

and validation control process, fulfill many government regulations, and follow tight standard procedures.

The progress of the medical product developments is not carried out in a linear way but in feedback loops and overlapping between the different phases. Basically, in the early stages, the prototypes are developed; then the preclinical tests allow the validation of designs through a group of tests for toxicity, biocompatibility, and security, which leads to several redesigns and adjustments to the device created. Next, with the improved designs, the clinical tests allow the analysis of these issues in a more accurate way, leading to more redesign that assures the safety and reliability, and then the results of these tests have to pass through a regulatory mechanism in accordance with the requirements established by regulatory agencies in each country. Occasionally, product evaluations and modifications continue to occur even after a product reaches the market [6].

From an engineering perspective more focused on product design, El-Haik and Mekki [7] propose the linear life cycle model shown in Figure 2.2. The life cycle is divided into 10 phases from ideation to disposal.

The first to fifth phases are similar to those phases used for developing conventional products. In fact, the more conventional tools can be applied in each one of them. First, the idea is created using techniques such as research and development ideation, benchmarking, technology road maps, and/or multigenerational plans. Second, the key functional requirements and all the constraints from customers and regulatory bodies in the countries where the device will be sold are identified, as well as any people in contact with the product (stakeholder). The quality function deployment (QFD), axiomatic design, or functional analysis is used in this phase. Third, several

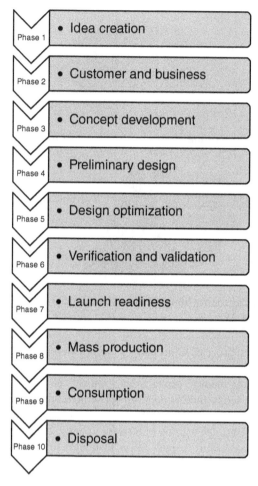

Figure 2.2 Linear life cycle model of medical device according to El-Haik and Mekki [7].

concepts are proposed. TRIZ theory (also known as Theory of Inventive Problem Solving) and the morphological matrix can be used in this proposal, whereas Pugh selection matrix helps in selecting the best concepts created. In the fourth phase, most design parameters that satisfy the functional requirements and constraints are defined leading to the design structure. Final design is adjusted defining all the details and using techniques such as Design for X (DFX), Failure mode and effect analysis (FMEA), and Manufacturing process selection (MPS) to ensure that the design cannot be used in a way that was not intended, to guarantee the manufacturability and to detect any possible mistake before arriving to the market. The result of detailed design is the production of a prototype form.

From phase 6, the processes are more specific for medical device development. In phase 6, the verification and validation procedures are developed, and the prototype is created. It is tested thoroughly and results in making final adjustments to ensure that the product matches all stakeholders' requirements and constraints. Next, in the launching phase, all production infrastructure and other resources are assessed. Here, the standard operating procedures, the training procedures of the personnel, and other special measures required are documented and identified. In the production phase, the manufacturing processes should be able to manufacture without defects, fulfilling all the quality requirements established in phase 6. The quality methods used in this stage include statistical process control, Taguchi method, robust design, production troubleshooting, and diagnosis methods. Finally, in the last two stages, the product is consumed, retired, and eliminated. Suitable after-sale services will help to keep the design in use by repairing defective units and extending its life cycle.

2.1.2 Medical Device Development Process

It is widely known that an effective new product development (NPD) affects strongly the likelihood of their success in the market [8]. Consequently, it is important to integrate the product development models available in the literature for developing these new products.

In general, these models state the set of activities to be followed to ensure the success of the final design, control the flow of information by ensuring that the right people use the right information at the right time, and moreover detail the set of techniques to be used to guarantee that decisions are taken as appropriate as possible. In addition, it is generally known that the use of these standardized models of product development processes (PDP) reduce development time and the final product cost. And in that sense, the medical device sector is not different [1].

In the literature, several models are proposed for the development of medical devices. In 2009, Pietzsch et al. [9] proposed the stage-gate process from concept to commercialization specifically for the development of medical devices (Figure 2.3). The model was developed based on in-depth interviews of experts actively involved with the development, regulation, and use of medical devices. The stage-gate process includes the following five phases: (1) *initiation*—opportunity and risk analysis, (2) *formulation*—concept and feasibility, (3) *design and development*—verification and validation, (4) *final validation*—product launch preparation, and (5) product launch and postlaunch assessment. In each phase, the essential activities are defined to each functional group related to the product. Due to the complexity of this process, Shluzas and Pietzsch detailed the interaction between the phases and the activities [10].

In 2013, Medina et al. [8] developed a generic model which integrates product development, regulation, standardization and patenting. The model is divided into five clusters according to information type, which are product development and introduction process, medical specifications, US FDA, standards, and patents. The main activities and relations are detailed for each cluster.

In 2012, Das and Almonor [11] proposed an attribute-driven concurrent engineering (ADCE) process for the development of medical devices based on concurrent engineering approaches. This model identifies the main activities in biomedical product development with a flowchart in a simplified way. In addition, it proposes

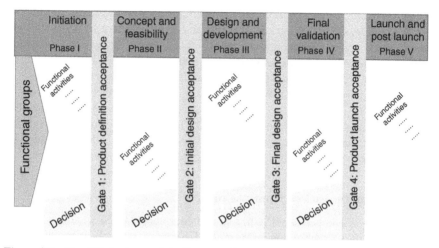

Figure 2.3 Simplified schema of the linear medical device development model according to Pietzsch et al. [9].

the formulation of attribute-driven specification (ADS) to consider several types of knowledge such as engineering, medical, marketing, quality, regulatory issues, and manufacturing. Three types of attributes are proposed: customer, functional, regulatory, which have to be documented and quantified. Also, ADS-driven design control is proposed for assuring the fulfilling of the requirements.

Neelamkavil et al. [12] developed a methodology that aimed on the traceability of all the information evolved during the development of medical devices—from customer needs to manufacturing and testing. The methodology aims to help users capture, manage, and reuse engineering knowledge during the various stages of product development.

2.1.3 Medical Devices' Design Process

The medical device design process involves issues from the engineering, business and medical fields, and integrate matters from government regulatory (domestic and international) and certification agencies. To guarantee the safety and effectiveness of the medical products, the use of design controls, which provide a methodology of assessing the design process through different phases of the product development, has been adopted. According to Gilman et al. [13], design controls are a set of methodical procedures to be followed during a design process to assure that the resultant product is safe, effective, and can be successful in a competitive marketplace.

The design process of biomedical products pass through the same phases of a conventional product design: concept, development, manufacturing, and distribution. However, a verification test plan must be defined during the product development phase, and a product validation test plan is to be applied during the manufacturing phase [13].

Validation and *verification* are wide terms that are defined in a different way depending on the regulatory bodies. According to Alexander [14] and Alexander and Bishop [15]: *"validation is concerned with demonstrating the consistency and completeness of a design with respect to the initial ideas of what the system should do and verification is concerned with ensuring that, as the design and implementation develop, the output from each phase fulfills the requirements specified in the output of the previous phase."* In order to clarify the official definitions, the US FDA provides the waterfall model in its document entitled *Design Control Guidance for Medical Device Manufacturers* that provides a more useful schema of validation and verification (Figure 2.4).

The waterfall model indicates that verification assures that the requirements established in the design input are fulfilled in the design output and implies a detailed examination of several issues of a design at various stages of product development. Validation is a much more involved process than verification. Design validation is a cumulative summation of all efforts, including design verification, and extends to the assessment to address whether devices produced in accordance with the design actually satisfy user needs and intended uses.

The view provided by Alexander and Bishop [15] about verification and validation is really interesting, and they view these concepts answering the following questions:

- *Verification*: are we building the thing right?
- *Validation*: have we built the right thing?

Considering the complexity and broadness of validation, Alexander [14] proposes a practical approach aimed at making devices easier and economical to validate. The approach is structured by a model of design for validation (DFV-V model), which shows the main steps to convert user needs to a physical medical device. These steps are "device design," "process design," and "production development," and each one

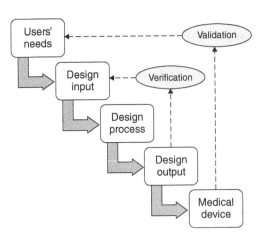

Figure 2.4   Schema of waterfall model [16].

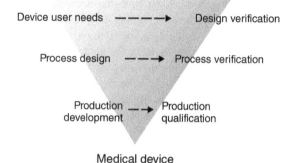

Medical device

Figure 2.5 Schema of design for validation model (DFV-V model) according to Alexander and Bishop [15].

is related to the corresponding procedure of validation and verification by means of verification requirements (Figure 2.5).

2.2 CASE STUDY

This section presents the application of medical product development process applied in two case studies: scapholunate prosthesis and stent tracheal. In both cases, a preliminary prototype was manufactured, although preclinical testing has not been carried out yet. Several conventional design techniques have been applied in each case.

2.2.1 Scapholunate Interosseous Ligament

Scapholunate instability can cause severe pain, swelling, and reduction of motion due to a break in one or more intercarpal ligaments [17, 18], the scapholunate interosseous ligament (SLIL) being the most affected one. Several surgical approaches have been recommended to manage SLIL chronic tears, including scaphotrapeziotrapezoidal and scaphocapitate fusion, capsulodesis, ligament repair with dorsal capsulodesis, tenodesis, and bone-ligament-bone reconstruction [19]. However, most of these techniques imply open surgery leading to extended recovery times, relevant scarring, and very often stiffness in the wrist joint [20].

Arthroscopy is commonly used as a minimally invasive technique to accurately detect and diagnose carpal ligament tears and allows the direct observation of the intrinsic and extrinsic carpal ligaments under static and dynamic instability conditions. Focusing on arthroscopy surgery, the combination of arthroscopy techniques and prosthesis could be a good solution to manage it [21, 22].

A new prosthesis to replace the functionality of the SLIL was designed and manufactured. The design process was assessed by expert surgeons in wrist arthroscopy, and the prototype manufactured allowed the surgeons to analyze the advantages and drawbacks of the adopted solution. The knowledge acquired by applying formal

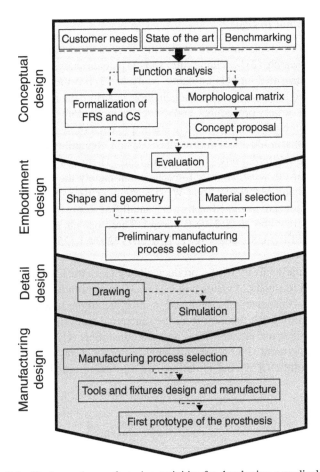

Figure 2.6   Design and manufacturing activities for developing a medical device.

design methods and the traceability of this knowledge through different stages of product development make new developments, design process control, and validation easier.

The methodology used to design the prosthesis is based on the systematic design methodology proposed by Phal et al., [23], one of the most common procedures to design conventional products. Figure 2.6 shows the activities carried out to gather the design product and the prototype. In the design process, first, the main design concepts were proposed and evaluated with expert doctors, and the final selection was refined. Next, the shape and the geometry were refined in embodiment design, and the material and preliminary manufacturing processes were selected. Finally, in the detailed design phase, the drawings were completed.

To develop the prototype, first, the final process for manufacturing the prosthesis was chosen, next the tools and fixtures were designed and manufactured, and finally a first prototype of the prosthesis was produced.

2.2.2 Conceptual Design

The methods for solving the SLIL instability were studied in three different ways: (1) scientific research, (2) benchmarking, and (3) patents analysis. Scientific research trends can be classified into (1) implants' analysis, (2) prosthesis design for solving problems related to the wrist, (3) wrist behavior regarding the movements and forces of the wrist, (4) surgery procedures, and (5) design methodologies applied in medical devices. The more relevant papers were studied in each of these fields to capture all the knowledge that could affect the prosthesis design. Regarding benchmarking of the market, there are different types of wrist implants. There is a group used to replace several bones of the wrist, another to replace completely the wrist (E-MOTION™ Total Wrist System, Maestro™ Total Wrist System, or Biax™ Total Wrist System), and other implants are designed to fix very small fragments of specific bones (Herbert™ Bone Screw and HCS 2.4/3.0 Screw). Three patents are available to repair or replace the SLIL: US0306480A1 [24], US0177291A1 [25], and US0076504A1 [26].

The functional analysis of the product allows to capture all the functions that the product has to satisfy. As shown in Figure 2.7, a minimum of three levels in the hierarchical analysis have to be deployed to integrate all the functions that the prosthesis has to fulfill. Each function from the functional analysis has to be formalized by the functional requirements (FRs) and the constraints (Cs) related to it [27]. A functional requirement (FR) represents what the product has to do independent of any possible solution [23] and is defined by the function and the qualifiers [27];

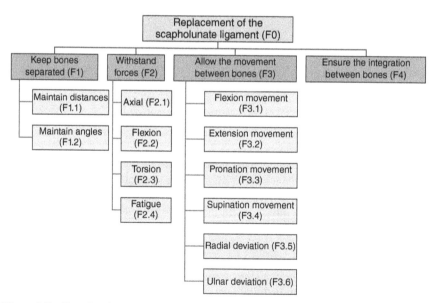

Figure 2.7 Functional analysis of the prosthesis to replace the scapholunate interosseous ligament (SLIL).

TABLE 2.1   Identification of Functional Requirements (FRs) and Design Constraints (Cs) for New Prosthesis

Functional Requirements		
Code	Function	Functional Constraint
FR11	To keep the scaphoid and lunate bones separated at distance d	$1 - 2\,\text{mm} \leq d \leq 5\,\text{mm}$
FR12	To allow relative motion between the scaphoid and lunate bones	Flexion, extension, deviation
FR21	To support axial loads	$28 \pm 8, 6\,\text{N}$ [28]
FR31	To allow flexion movement	76° [29]
Design Constraints		
Code	Description	Quantifier
Cs1	Implant biocompatible materials	List
Cs2	Implant maximum dimensions	Diameter: 4 mm, length: 25 mm
Cs4	Easy to insert	Subjective
Cs5	Easy to remove	Subjective
Cs6	Maximum internal hole diameter	Internal diameter 1 mm
Cs7	Self-rotating thread	Pitch (mm)

see Table 2.1. The qualifiers limit the possible design solutions and are represented by the constraints (Cs). There are two kinds of constraints [27]: functional and design (Table 2.1). Functional constraints limit directly the function or the action, for example, the range of distance that the scaphoid and lunate can be separated (FR12 in Table 2.1). Design constraints are the limitations related directly to the component's physical definition (e.g., the maximum dimensions of the prosthesis, defined in Cs2) or the ones that come from the conditions where the action is being carried out (e.g., the compatibility material properties specified in Cs1). At least one functional constraint has to be defined for each FR, and each FR can be limited by more than one design constraint [27].

The information to quantify the functional requirements and constraints was obtained from different sources. For example, those quantifiers related to loads and movements (FR21 and FR31) were obtained from the scientific literature, whereas those related to anatomy or surgery process came from the doctors' expertise (FR11, FR12, Cs2, Cs6, or Cs7). When the quantitative value is not available, the FRs cannot be measured or validated; consequently, it is difficult to know whether it is being fulfilled or not.

Taking the functions from the functional analyses, a morphological matrix was developed to create possible solutions. Therefore, possible physical solutions were identified for each function of the functional analysis. Figure 2.8 shows a synthesis of the results.

The combination among different physical solutions leads to three main design concepts: the compact system (concept 1), the ball-and socket-ball system (concept 2)

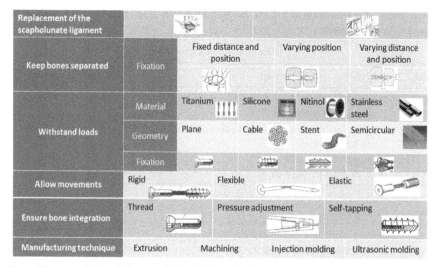

Figure 2.8 Synthesis of the morphological matrix. (*See color plate section for the color representation of this figure.*)

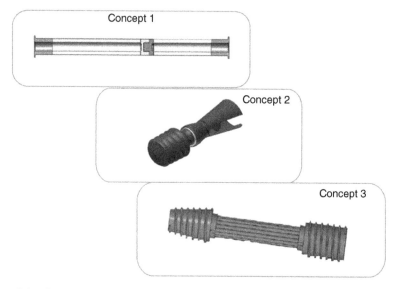

Figure 2.9 Conceptual design: compact system (concept 1) and ball-and socket-ball system (concept 2) and wired system (concept 3). (*See color plate section for the color representation of this figure.*)

and the wire system; see Figure 2.9. The compact system allows varying the position and the distance between the two bones due to the material elastic properties, and the bone integration is done by rounded ends. The ball-and-socket-ball system allows only to vary the position, keeping the distance fixed; the geometry is quite variable and the integration with the bone is done by a press-fit system. The wire system is very flexible, as two braided wires placed in opposite directions allow recovering the starting position. Three main reasons were given by the doctors to choose the compact system: (1) the movements are less limited due to the elasticity of the material; (2) it is easier to insert and remove; and (3) it is a novel idea to solve this problem. From an engineering point of view, the first concept contains fewer parts and, therefore, the manufacturing process will be more manageable.

2.2.3   Embodiment Design

The compact system is divided into three different parts: scaphoid part (rigid), lunate part (rigid), and elastic part.

The knowledge of the doctor allows refining it in three different stages, which are shown in Figure 2.10. First, the intermediate joint was changed to the end of the device, as the bones could be damaged by this union. The ends were changed to cylindrical shape with a little conic degree, and threads were added to improve the bone fixation and integration. The union between the rigid and elastic parts was shaped by rectangular slots to improve the fixation among them. Inside the prosthesis, a hole was introduced. This hole has a hexagonal end to transmit the rotational movement during the insertion of the prosthesis into the bones. Second, the conic shape from the scaphoid was changed to cylindrical, as it is enough to assure the assembly in the

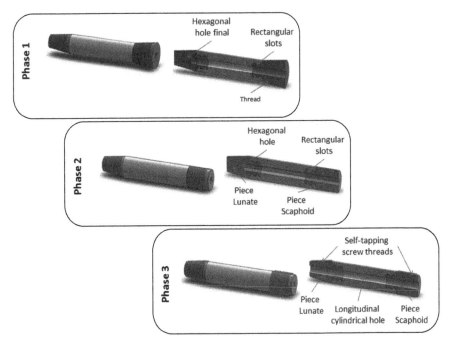

Figure 2.10   Evolution of the compact system. (*See color plate section for the color representation of this figure.*)

bone. The internal hole is completely hexagonal to improve the transmission of the torque while the prosthesis is being inserted during the surgery process. And finally, the rigid ends were adapted with self-tapping screws threads. A circular hole was created in the prosthesis using a Kirschner needle (1 mm diameter) to guide the device during surgery. And finally, some essential dimensions were also adjusted, such as diameter and length.

2.2.4   Detailed Design

The final design was refined to obtain the final drawing with all the dimensions, materials, and details to be manufactured. The prosthesis length is less than 20 mm, and its maximum diameter is 3 mm (Figure 2.11a). The material chosen for the rigid parts was biocompatible titanium, whereas for the elastic part, biocompatible liquid silicon rubber (LSR) was chosen. Finally, in the design process, the device was inserted in a real situation using images processed from a CT scan (Figure 2.11b).

2.2.5   Manufacturing a Prototype

The task to manufacture the design is summarized in Figure 2.12. First, the scaphoid and lunate parts were manufactured using an Okuma lathe available in our laboratory. Considering the capabilities of this machine, several tests related to manufacturing strategies, tools, and materials were needed to obtain the final parts. Next, the mold to pour the silicone that contains the rigid parts was also designed and manufactured. Two different design versions were designed: rectangular and cylindrical. Although the rectangular mold appeared to be more suitable for the pouring process, the cylindrical one was chosen due to the easiness in making it in our lab. The external manufacturing of the mold and other parts had increased the development time and product cost, as being the first prototype, this was not desirable.

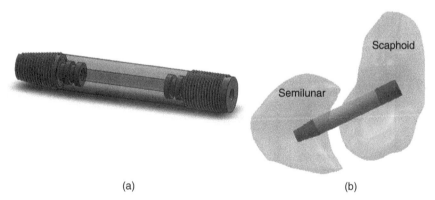

(a)   (b)

Figure 2.11   Final scapholunate prosthesis (a) and its insertion in the bones (b). (*See color plate section for the color representation of this figure.*)

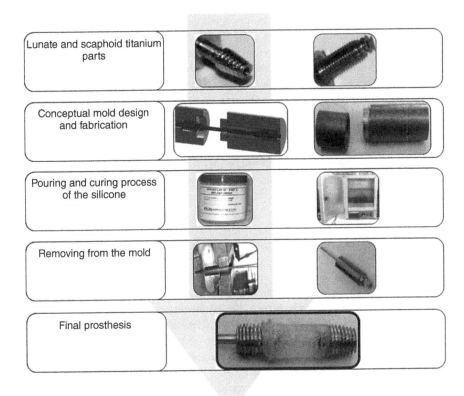

Figure 2.12 Manufacturing steps. (*See color plate section for the color representation of this figure.*)

In the pouring process, first, the titanium parts and the Kirschner needle were introduced in the mold; second, the silicone was poured; third, the mold was closed; and finally, it was introduced in an oven to cure the silicone material. In the final task, the mold was removed carefully to avoid damaging the final prosthesis.

As mentioned earlier in this section, this first prototype was manufactured using machines available in our lab, which are generally focused on meso-manufacturing. Consequently, several manufacturing tests were needed, and the final quality of the prosthesis was not as high as expected.

To conclude, according to the doctors, this novel device has the following advantages in managing the instability of the SLIL: the combination of rigid and elastic materials to allow the variation in the relative distance and the position between the bones, easiness in insertion, and the simplicity of the design (a single piece to avoid assembling during the surgery process).

In addition, the application of the rigorous and systematic design methodology allowed capturing all the information and knowledge related to this case study, which could be used for proposing other designs quickly and assuring that the customer's needs are satisfied. Although it is a good design approach, it should be improved from a material point of view. According to the doctors, these kinds of silicones are not suitable in this case, and the union cannot be guaranteed.

In the manufacturing process, micro-manufacturing machines should be used for manufacturing the prototype. New technologies, such as ultrasonic molding, extrusion, or injection for the final assembly and micromachining for the rigid parts and the mold, should be analyzed. Consequently, the elastic material could be better processed and its union with the rigid parts could be better achieved.

Furthermore, a Finite Element Method (FEM) would be required to know the exact behavior of this prosthesis in the bones from a movements and loads point of view.

2.2.6   Tracheal Stent

Tracheobronchial stenosis is the obstruction of the tracheal air passage that can be caused by different diseases, such as malignant or benign tumors, tracheobronchomalacia, extrinsic compression, and postintubation tracheal injuries. One palliative treatment available for patients with tracheal obstruction is stenting [29] [30]. This form of therapy provides immediate relief from life-threatening conditions and significantly improves the patient's quality of life. Moreover, stenting may also function as a bridge until further curative treatment can be pursued. Nowadays, there are several different tracheal stents, but there is no perfect shape and no easy and quick way of manufacturing these.

This case study presents a methodological model based on a set of design techniques to design a functional and innovative tracheal stent, as well as the implementation of a parameterized tool that allows customizing the final design to suit quickly and accurately the patient's parameters. Some of the steps followed during the study are

- research and benchmarking of tracheal stents,
- design development,
- parameterization of final stent shape for customization,
- selection of materials,
- selection of the manufacturing process,
- manufacturing of new tracheal stent prototype.

As mentioned in this chapter, design methodologies include useful tools and techniques for developing new products or improving them. In this case study, three different design tools and one innovation algorithm were used, which are attribute listing, QFD, and TRIZ.

> Attribute listing focuses on the attributes of an object, seeing how each attribute could be improved. Attributes are parts, properties, qualities, or design elements such as dimensions, weight, style, and flexibility.

QFD is a systematic process that helps planning the development of new products or improving them, by fully understanding the necessities and requirements of the client [31]. QFD is a useful tool to reinforce the conceptual design stage, which helps in visualizing many attributes of the product and summarizes them in a set of graphics known as "Quality tables," where information related to the client is collected [32].

The attributes coming from the attribute listing can be translated into WHATs in the QFD matrix, and next, the technical requirements of each one of them can be established by means of HOWs in the QFD [31]. The technical requirements or HOWs of the QFD represent the engineering characteristics and describe the product from a designer/engineer point of view.

The relationship between the WHATs and HOWs is the main body of the QFD, and its purpose is to translate the customer needs into the technical character-istics, which allows design of the final product. The level of interrelationship discerned is weighted usually on a four-point scale (high, medium, low, none), and a symbol representing this level of interrelationship is entered into the matrix cell. The target values is the final section of the QFD to be completed, and it summarizes the conclusions drawn from the data contained in the entire matrix and the team's discussions. These are a set of engineering target values to be met by the new product.

TRIZ theory (Teoriya Resheniya Izobretatelskikh Zadatch, in Russian) is an inno-vation algorithm for inventive problem solving based on science and technol-ogy, requiring a high degree of creativity and inventiveness [32]. By analyzing more than 1.5 million patents worldwide, the developers of TRIZ found that all technical systems that evolve are governed by objective laws. These laws reveal that, during the evolution of a technical system, improvement of any part of that system having already reached its pinnacle of functional performance will lead to conflict with another part, that is, when technical contradictions appear. The tools to overcome technical contradictions are called principles. Principles are generic suggestions for performing an action to, and within, a technical sys-tem. There are 40 principles in total within the TRIZ algorithm, and they allow the development of numerous solution concepts for every technical problem. Implementing a chosen concept still remains the work of an engineer [28]. The intention of TRIZ is to minimize the number of trials and errors, not to provide a magical solution.

2.2.7 Conceptual Design

The design tools described have features that enable their integration making the conceptual design easier, thus more efficient. However, first of all research and bench-marking were carried out.

Based on an extensive literature research, a benchmarking analysis for stents, and information retrieved from physicians and patients, a list of desirable attributes for the stent was created. Many of these attributes are oriented toward a better fit of the stent to the patient's trachea to avoid migration, fast and low-cost production of the stent, as

TABLE 2.2   Attribute List of Stent

Attribute	Requirements
Easy to insert and remove	An elastic modulus of 1–15 MPa is required to insert and remove a stent
Dynamic	Ability to respond to cough pressures (7500 Pa) with reversible reduction of cross-sectional area of the stent. The wall near muscular tracheal tissue must have a displacement during coughing ≤4 mm (avoid migration) Shore $A = 70$–75
Low rugosity	Rugosity level $R_a = 200$ nm
Biocompatible	As stated by international regulations, the material must comply with ISO 10993-1 US FDA approved for long-term implantation The material must have a thermal stability at 35–45 °C ISO 10993-18:2005: Chemical stability, it must not react with the environment (air, body cells, blood)
Low cost	The final cost has to be ≤$1100 Average silicone stent cost is $250–600, plus using rigid bronchoscope could increase the net reimbursement by $1000
Hydrophobic	Hydrophobic = 99–130° contact angle
Radiopacity	Inability of X-ray or chemotherapy to pass through a material: $I < I_0$ $I = I_0 e^{-\mu x}$ $I =$ original intensity $I_0 =$ transmitted intensity $\mu =$ attenuation coefficient $x =$ thickness of material
Fast manufacturing	Lead time: ≤10 h

well as traits such as radiopacity. The material or materials used to fabricate the stent must be bioimplantable to avoid any adverse reactions. Table 2.2 shows the summary of these requirements.

Once the attribute list is made, it is easy to find the main necessities (WHATs) and establish functional requirements (HOWs) to each one of them [31]. A total of 10 necessities were identified; each one has a functional requirement that must be translated into a real numerical target (Figure 2.13a). For example, if the necessity is "Flexible" then one of the functional requirements can be "Reduce wall thickness," so the target is "Less than 0.5 mm." Each necessity has a strong, moderate, weak, or no relation with all requirements. The display of WHATs and HOWs in the QFD is shown in Figure 2.13.

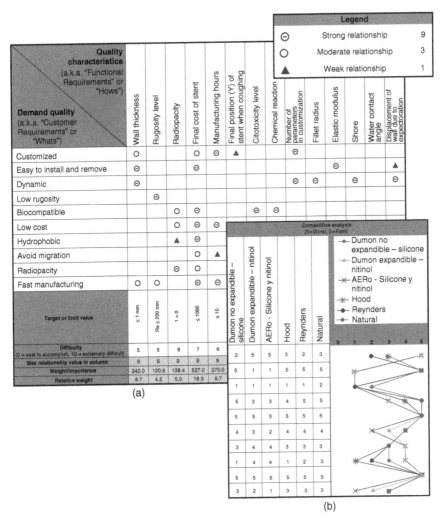

Figure 2.13 (a) Display of WHATs and HOWs in QFD and (b) QFD competitive analysis.

In the competitive analysis section of the QFD, there is a comparison between six commercial stent designs (Figure 2.13b). The competitive analysis helps grading the commercial stents based on how well they cover each necessity—5 being the best grade and 1 being the worst. As we can see in Figure 2.5, the designs that cover most of the necessities are the natural stent and the Dumon nonexpandable made of silicone. Both designs have the lowest grade in "Dynamic" and "Migration;" the fact that both are made of silicone makes them flexible but not enough for changing shape when coughing, and they slip easily. On the other hand, they are better, compared to expandable stents, because they are less expensive, easier to insert, and have excellent biocompatibility.

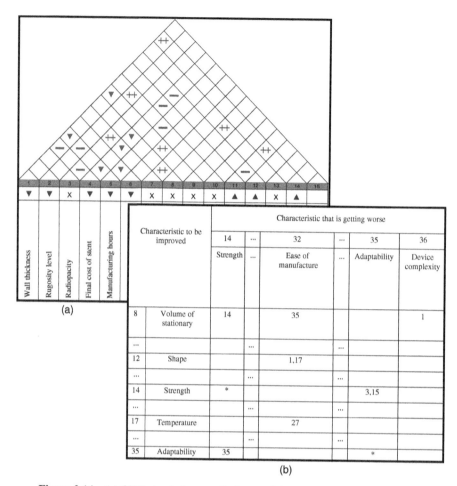

Figure 2.14 (a) QFD physical contradictions and (b) TRIZ contradiction matrix.

The QFD roof summarizes the correlations found among the characteristics (HOWs), which can be positive or negative (Figure 2.14a). A strong positive correlation occurs when a characteristic is improved and a desired effect occurs on another characteristic. Otherwise, a correlation is negative when a nondesired effect occurs on another characteristic. Eight strong positive correlations, seven negative correlations, and six strong negative correlations were found in the QFD roof; see Figure 2.14a.

TRIZ principles are applied to the negative and strong negative correlations identified to solve each contradiction (Figure 2.14b). Some examples are as follows:

- Principle 1 (*Segmentation*: divide the object in independent parts) translated in our case as "Segment the shape to avoid migration."

- Principle 14 (*Curvature*: instead of using rectilinear surface or form use curvilinear ones).
- Principle 15 (*Dynamics*: if the manufacturing process is rigid, make it adaptive, divide the object into parts capable of movement between each other) translated in our case as "Adapt the additive manufacturing technology for an effective processing of the material" and "Add edges to the stent with different shapes."
- Principle 3 (*Local Quality*: Change the object from uniform to nonuniform).
- Principle 1 (*Segmentation*: divide the object in different parts) translated in our case as "Add different flares or studs that can move easily to avoid migration" and "Add other objects such as springs to absorb expectoration forces."
- Principle 17 (*Another Dimension*: use another side of a given area) translated in our case as "Add flares or studs only in a certain part of the stent."
- Principle 35 (Change the concentration or consistency, change the degree of flexibility, change the processing or curing temperature).
- The material must have the following characteristics: approved for long-term implantation, with high elastic modulus and Shore A as well.
- Principle 35 (*Parameter Changes*: change the processing temperature, change the material's physical state) translated in our case as "If the processing temperature is difficult to achieve, then treat the material later or treat the material in one state and then change it."
- Principle 27 (*Cheap Short Living Objects*: use disposable objects) translated in our case as "If the material cannot be processed directly by the manufacturing technology, change the approach, molds for example."

2.2.8   Embodiment Design and Detail Design

The TRIZ engineering principles were divided into solutions for shape, material selection, and manufacturing technology and used to provide solutions for each area. These solutions are explained in the following.

- *Solutions for Shape*   By means of brainstorming, many sketches were drawn following these TRIZ solutions. As a result, all sketches had three things in common: D shape of stent resembling trachea's natural geometry, reduced amount of material in muscular wall, and rounded fillets in every edge or ring. There were 11 models after design brainstorming (see Figure 2.15), and all designs underwent a voting session.

  The grading was done by three experts in manufacturing, four designers, and four surgeons, and the model with highest grading is shown in Figure 2.16. Here, the parameters for customization for this model are shown: outside diameter, OD; inside diameter, ID; membrane thickness, m_t; wall thickness, w_t; length of stent, L; and distance between rings, l.
- *Solutions for Material: Material Selection*   The main requirements include biocompatibility and permanent implant grade. Therefore, a liquid long-term

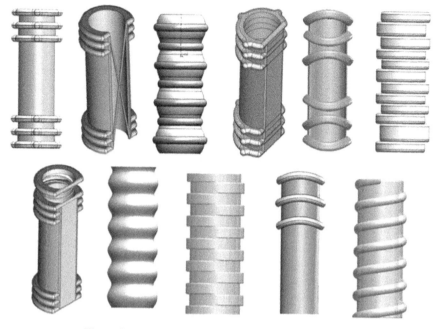

Figure 2.15   Geometry solutions using TRIZ principles.

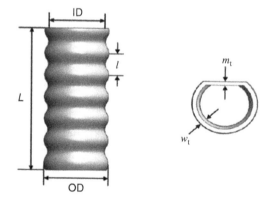

Figure 2.16   Customization parameters of final stent design.

implant-grade platinum-cured silicone rubber (LSR) was selected studying a wide list of materials.

- *Solutions for Manufacturing: Manufacturing Process Selection*   The QFD shows that the stents have to be manufactured with a rapid prototyping technology in order to satisfy the necessities: low cost and less manufacturing

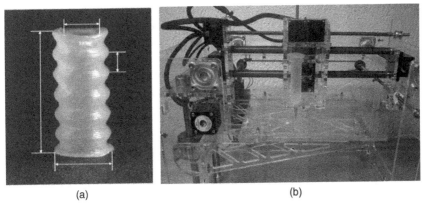

(a)   (b)

Figure 2.17   Stent (a) directly manufactured using fab@home (b), a fused deposition modeling machine.

hours. Because patients with tracheobronchial cancer have low life expectancy, immediate operations are frequent, and hospitals run out of standard stents most of the time, which makes it difficult to get new stents on an urgent basis. For instance, implant models can be produced and be ready for evaluation within hours using additive manufacturing technologies such as stereolithography, selective laser melting (SLM), selective laser sintering (SLS), fused deposition modeling (FDM), electron beam melting (EBM), or 3D printing. These methods and their related technologies are being widely used to manufacture customized implants and prosthetics [30]. FDM technologies were selected.

2.2.9   Manufacturing a Prototype

For the tracheal stent design presented, several manufacturing approaches have been explored, all based on additive manufacturing (AM). First, to prove the manufacturing concept, a nonimplantable silicone was used with direct manufacturing using a commercial silicone and the fab@home, a low-cost FDM machine (Figure 2.17). To prove the manufacturing concept, a nonimplantable silicone, was used.

However, long-term implant polymers, such as medical silicones, change phase rapidly, and their processing by FDM technologies is very difficult. The most suitable processing technology for liquid silicone rubber materials is injection molding, considering an adapted mold to obtain this part and the technology to inject this specific material.

2.3   CONCLUSIONS

The combination of attribute listing, QFD, and TRIZ presented herein were used to guide the development of a new stent design suitable to palliate an airway obstruction.

The major results of this design methodology were the identification of the engineering requirements and generation of an innovative stent geometry, and selection of material and the necessary technology for its manufacturing. As a difficult task, some of the customer requirements such as biocompatibility, fast delivery, and lack of migration were considered. A wavy D-shape stent was created, intended for avoiding migration and a posterior wall thickness, which may potentiate the dynamism of the trachea and prevent its obstruction. Also, the material selection and its manufacturing process were easier to establish by arranging all the requirements in the QFD and by applying the TRIZ concepts to the physical contradictions.

The stent prototype fabrication was possible with the use of additive manufacturing technologies; however, the selected biocompatible silicone could not be handled directly with these technologies. A rapid tooling approach was selected and several mold iterations were manufactured and tested using FDM low-cost machines. The results of the molds' trials suggest that a better surface finish in the cavity and the addition of venting channels in the parting line can improve the mold's performance.

REFERENCES

[1] Santos ICT, Gazelle GS, Rocha LA and Tavares JMRS. Medical device specificities: opportunities for a dedicated product development methodology. Expert Rev Med Devices 2012;9(3):299–312.

[2] Hubka V, Eder WE. *Design Science: Introduction to Needs, Scope and Organization of Engineering Design Knowledge.* Berlin, New York: Springer; 1996.

[3] Suh NP. *Axiomatic Design: Advances and Applications.* New York, Oxford: Oxford University Press; 2001.

[4] Hrgarek N. Integrating six sigma into a quality management system in the medical device industry. J Inf Organ Sci 2009;33(1):1–12.

[5] Santos ICT, Tavares JMRS. Additional peculiarities of medical devices that should be considered in their development process. Expert Rev Med Devices 2013;10(3):411–420. http://www.ncbi.nlm.nih.gov/pubmed/23668711.

[6] CDRH, and US FDA. The Journal of Infectious Diseases *CDRH Innovation Initiative.* Suppl 209; http://www.ncbi.nlm.nih.gov/pubmed/24872391; 2011.

[7] El-Haik BS and Mekki MS. *Medical device design for Six Sigma. A Road Map for Safety and Effectiveness.* John Wiley & Sons, Inc. EEUU; 2008

[8] Medina LA, Kremer GEO, Wysk RA. Supporting medical device development: a standard product design process model. J Eng Des 2013;24(2):83–119. http://www.tandfonline.com/doi/abs/10.1080/09544828.2012.676635 (May 23, 2014).

[9] Pietzsch JB et al. Stage-gate process for the development of medical devices. J Med Devices 2009;3(2):1–15.

[10] Shluzas LA, Pietzsch JB. The interactive nature of medical device design. In: *International Conference on Engineering Design.* (ICED 09), Vol. 1, Design Processes, Palo Alto, CA, USA, 2009, p 85–96.

[11] Das SK, Almonor JB. A concurrent engineering approach for the development of medical devices. Int J Computer Integr Manuf 2000;13(2):139–147.

[12] Neelamkavil J, Pardasani A, Kernahan M, Ng C. Traceability in Medical Devices Design & Manufacturing. Proceedings of the Canadian Design Engineering Network (CDEN) Conference, Kaninaskis, Alberta, July 18–20, 2005.

[13] Gilman, BL, Brewer JE, Kroll MW. Medical device design process. Conference proceedings: Annual International Conference of the IEEE Engineering in Medicine and Biology Society. IEEE Engineering in Medicine and Biology Society. Conference 2009: pp. 5609–5612. http://www.ncbi.nlm.nih.gov/pubmed/19964397; 2009.

[14] Alexander K. A validation model for the medical devices industry. J Eng Des 2002;13(3):197–204.

[15] Alexander K, Bishop D. *Good Design Practice for Medical Devices and Equipment – A Framework*. University of Cambridge Engineering Design Centre, Cambridge, UK; 2001.

[16] US Food and Drug Administration. Center for Devices and Radiological Health. Design control guidance for medical device manufacturers. Available at: http://www.fda.gov/downloads/MedicalDevices/.../ucm070642.pdf; 1997.

[17] Slater RR, Szabo MR, Bay KB, Laubach J. Dorsal intercarpal ligament capsulodesis for scapholunate dissociation: biomechanical analysis in a cadaver model. J Hand Surg 1999;24:232.

[18] Zubairy AI, Jones WA. Scapholunate fusion in chronic symptomatic scapholunate instability. J Hand Surg 2003;28:311.

[19] Whitty LA, Moran SL. Modified Brunelli Tenodesis for the Treatment of Scapholunate Instability. Ch 49 In: *Fractures and Injuries of the Distal Radius and Carpus*. Elsevier; 2009. p 481.

[20] Kalainov M, Cohen MS. Treatment of traumatic scapholunate dissociation. Journal Hand Surgery Am 2009;34:1317.

[21] Villanova JF, Del Pino JG. *Wrist Arthroscopy*. Elsevier; 2011.

[22] Kozin SH. The role of arthroscopy in scapholunate instability. Hand Clinics 1999;15:435.

[23] Pahl G, Beitz W, Wallace K, Blessing L, Bauert F. *Engineering Design: A Systematic Approach*. London: Springer; 1996.

[24] Kenneth Mitchell Wilson. Scapholunate Fixation Implants and Methods of Use. Patent Pub. No. 0306480 A1. 2008 December 11.

[25] David G. Jensen and Steven P. Horst. Orthopedic Connector System. Pub. No. 0177291 A1. 2008 July 24.

[26] Michael G. McNamara and Avery B. Muñoz. Method and Device for Stabilizing Joints with Limited Axial Movement. Pub. No. 0076504 A1. 2010 March 25.

[27] Ferrer I, Rios J, Ciurana J, Garcia-Romeu ML. Methodology for capturing and formalizing DFM knowledge. Rob Comput Integr Manuf 2010;26:420.

[28] Altshuller GS. *Creativity as an Exact Science: The Theory of the Solution of Inventive Problems*. CRC Press; 1984.

[29] Mroz RM, Kordecki K, Kozlowski MD, Baniukiewicz A, Lewszuk A, Bondyra Z, Laudanski J, Dabrowski A, Chyczewska E. Severe respiratory distress caused by central airway obstruction treated with self-expandable metallic stents. J Physiol Pharmacol 2008;59:491–497.

[30] Stephens KE, Wood DE. Bronchoscopic management of central airway obstruction. J Thorac Cardiovasc Surg 2000;119(2):289–295.

[31] Otto K. Product Design, Techniques in Reverse Engineering and New Product Development. Prentice Hall; 2001.

[32] Téllez HA, Modelo del proceso de diseño conceptual: Integración de las metodologías QFD, Análisis funcional y TRIZ, Master thesis from Instituto Tecnológico y de Estudios Superiores de Monterrey, Monterrey, N.L, México; 1997.

3

FORMING APPLICATIONS

KAREN BAYLÓN
Department of Mechanical Engineering, Instituto Tecnológico y de Estudios Superiores de Monterrey, Campus Monterrey, Monterrey, Nuevo León, Mexico

ELISABETTA CERETTI AND CLAUDIO GIARDINI
Department of Mechanical and Industrial Engineering, University of Brescia, Brescia, Lombardy, Italy

MARIA LUISA GARCIA-ROMEU
Department of Mechanical Engineering and Industrial Construction, University of Girona, Girona, Catalonia, Spain

3.1 FORMING

Forming of metal is a well-established process, dating back to more than a thousand years; in particular, it has been used by blacksmiths for the manufacture of weapons, armor, fittings, and other bellicose instruments. The idea of plastically deforming metals into a desired shape makes good use of the material and can even enhance its performance. This fact was well known in the ancient Asian cultures; in the forging of samurai swords, for instance, only by mastering the craftsmanship of forming and using their knowledge of materials and empirical skills for the hardening procedure, the sword maker was able to create the ultimate weapon.

All sheet metal forming processes can be divided into two groups: cutting processes, that is, process determining the material breakage, including shearing, blanking, punching, notching, piercing, etc., and plastic deformation processes, including

Biomedical Devices: Design, Prototyping, and Manufacturing, First Edition.
Edited by Tuğrul Özel, Paolo Jorge Bártolo, Elisabetta Ceretti, Joaquim De Ciurana Gay,
Ciro Angel Rodriguez, and Jorge Vicente Lopes Da Silva.

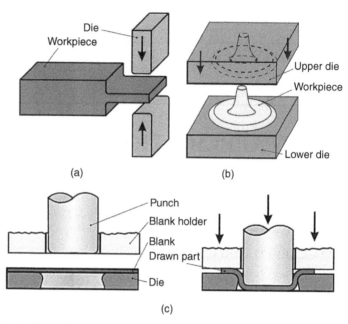

Figure 3.1   (a) Open-die forging; (b) closed-die forming; and (c) sheet stamping. *Source*: Schuler 1998 [2]. Reproduced with permission of Springer Berlin Heidelberg. (*See color plate section for the color representation of this figure.*)

bending, stretch forming, deep drawing, stamping, cogging bulk forming, and various other forming processes. The first group of processes involves cutting material by subjecting it to shear stresses usually between a punch and a die or between the blades of a shear. The punch and die may be of any shape, and the cutting contour may be open or closed. The second group involves partial or complete plastic deformation of the worked material [1] (Figure 3.1).

Sheet metal parts are usually made by forming materials at room temperature, although many sheet metal parts are formed in a hot condition because the material has a lower resistance to deformation when heated. Strips or blanks are often used as initial materials and are formed on presses using appropriate tools. The shape of a part generally corresponds to the shape of the tool [1].

Sheet metal forming processes are used for both serial and mass production. Their characteristics are high productivity, highly efficient use of material, easy servicing of machines, the ability to employ workers with relatively less basic skills, and other advantageous economic aspects. Parts made from sheet metal have many attractive qualities: good accuracy of dimension, adequate strength, light weight, and a broad range of possible dimensions, from miniature parts in electronics to the large parts of airplane structures [1].

Traditionally, sheet metals may be defined as metal having a thickness between 0.4 and 6 mm, while microsheet forming usually deals with sheet metals having a thickness usually less than 0.3 mm. Similar to conventional sheet metal forming, major material conversion mechanisms in microsheet forming include mainly shearing/cutting, bending, unbending, stretching, compressing, stress relaxation, and their combinations. In addition, the mechanical properties of the materials such as elasticity, plasticity, stress-strain relations, strain rate, work hardening, temperature effect, anisotropy, grain size, and residual stress are very important for understanding material deformation/separation mechanisms. The effects of grain sizes and orientations, and grain-boundary properties are especially significant in microsheet forming, considering their effects on the definition of the overall stress/strain relationships, sheared-section qualities, bending curvatures, springback phenomena, and stress relaxation. For a given microstructure, the effects are more significant, in terms of the relative ratios between the grain sizes and the strip thickness/feature sizes/part dimensions [3].

Today's bulk forming is used extensively in a wide range of industrial applications. A typical forming process requires multiple steps and comes in a wide range of variants. The industrial popularity of bulk forming process still relies on the theory of material flow and behavior, something that has not changed since the early days of forming. A well-functioning bulk forming system requires extensive knowledge of the strain hardening of the material and its flow behavior as well as friction and lubrication conditions (which involves tribology). Another important parameter in bulk forming is temperature. Hot operations take advantage of the decrease in flow stress, lower forces on the dies, and better material flow due to a working temperature higher than recrystallization temperature. This is fundamental when forming difficult to deform materials such as Ti, Mg, or Co alloys. For smaller deformation and when material hardening is expected, cold processes are more suitable.

Micro-bulk forming is the usage of bulk forming processes to manufacture micro-components such as hearing aids, drug delivery systems, and dental implants. Compared with more traditional micro-manufacturing processes, such as turning and milling, this process holds the potential of producing high-quality components, faster with no or only little material waste. The forming of small metal parts is not new, as industry has experienced this aspect for many years; however, challenges do arise when the sizes and features reduce to tens or hundreds of microns, or again the precision requirements for macro/miniature parts reduced to, for example, less than a few microns [4, 5].

Regarding existing hydroforming processes, a general distinction is to be drawn between the forming of tubular material, such as straight and bent tubes or profiles, and the forming of sheet material, for example, single or multiple sheets. Currently, tubular material is predominantly applied for the manufacture of hydroformed components. The principle of these hydroforming processes consists of basically placing at the beginning of the process the initial part into a die cavity, which corresponds to the final shape of the component, and the die is closed while the tube is internally pressurized by a liquid medium with internal pressure to carry out the expansion of the component. In addition, the tube ends are axially compressed by sealing punches to force material into the die cavity. The component is formed under the simultaneously controlled action of internal pressure and axial force. Water/oil emulsions are typically used to apply the internal pressure, which is usually increased to 120–400 MPa.

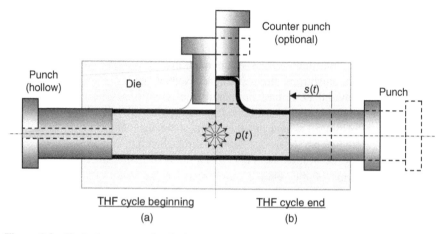

Figure 3.2 Typical sequence for T-shaped tube hydroforming process: (a) the tube in the hydroforming dies; (b) the produced part after hydroforming. Reproduced with permissions of Antonio Fiorentino.

The necessary amount of internal pressure is influenced significantly by the wall thickness and diameter of the component and the material strength and hardening, as well as the shape of it [3] (Figure 3.2).

Micro-hydroforming offers the potential to produce a wide range of products from the fields of medical engineering (e.g., needles and microtubes for drug delivery, micropipettes, tubular parts for endoscopes, or elements for surgical tools), microfluidics (e.g., components for microfluidic chips, elements for micro-dosage, or pipe connections and housings), and micro-mechatronics (e.g., shafts and elements for microactuators, components for microsensors or connection pins). However, certain design changes of such products with an adaptation to the micro-hydroforming process will be required to enable a failure-free and reliable production by hydroforming.

The medical sector is one important application of the forming manufacturing process, especially in the manufacture of micro-components. Here, the components are used as parts of dental implants, in spine fracture repair kits or even as elements in drug delivery systems.

3.2 TYPICAL PROCESS PARAMETERS

The principal processing parameters that affect the forming manufacturing process are discussed in the following sections.

3.2.1 Temperature

In forming operations, thermal energy is often supplied to the workpiece to increase its temperature; the workpiece needs to be at the proper working temperature in order to achieve the desired shape change and to have the proper microstructure for deformation reducing the forming force and the possibility of part breakage. This is particularly true for the production of orthopedic joints where Ti or Mg alloys are hot forged in closed dies. On the other hand, since dimensions of the workpiece for biomedical micro-components are under 1 mm or less, it is often preferred to work at room temperature, or higher but under the recrystallization temperature as well.

3.2.2 Flow Stress

It is found in many further researches that the flow stress increases with the increasing deformation. The most used parameter to describe the material behavior is the flow stress curve; this determines the forming force, the load on the tools, the local flow behavior (i.e., the material deformation), and thus the filling of the die cavities. As explained in the following section, with increasing miniaturization, the flow stress curves vary with the so-called size effect. This phenomenon occurs when in the thickness of the workpiece compasses just a few grains. The grain size effects are widely investigated, using the Hall–Petch equation. Lai and Peng [6] have proposed a model to describe the behavior of the material, in which the deformation in the material depends on the position of the grain related to the position in the cross-section of the workpiece; this model is called mixed material model. The Ling and Peng [6] model can be divided into two different terms: the size-dependent model σ_{dep} and the size-independent model σ_{ind}. This theory is supposed to model the behavior of the single crystal model and even the polycrystalline one.

$$\sigma(\varepsilon) = \sigma_{ind} + \sigma_{dep}$$

$$\sigma_{ind} = M\tau_R(\varepsilon) + \frac{k(\varepsilon)}{\sqrt{d}}$$

$$\sigma_{dep} = \eta\left(m\tau_R(\varepsilon) - M\tau_R(\varepsilon) - \frac{k(\varepsilon)}{\sqrt{d}}\right)$$

where $k(\varepsilon)$ is the locally intensified stress needed to propagate general yield across the polycrystal grain boundaries, M is the orientation factor for the polycrystalline model as well as m for the single crystal model, τ_R is the critical shear stress resolved for the single crystal, d is the grain size, and η is the size/scale factor.

This model has been validated by experimental and numerical results and used as reference in scientific articles, such as Wang et al., [7], concluding that the multiregion model, considering the grain orientation and the grain boundary, is reliable to be used in the simulation of micro-bulk forming process.

3.2.3 Strain

Plastic strain represents the permanent displacement of a body due to the application of an external force, and it is defined as

$$\varepsilon = \ln\frac{l}{l_0}$$

Stress and strain define the material flow stress and consequently the way in which material deforms and changes its shape.

3.2.4   Strain Rate

During the deformation process, the speed of the operation is usually measured by the strain rate, defined as the rate of change in strain of a material with respect to time:

$$\dot{\varepsilon} = \frac{\mathrm{d}\varepsilon}{\mathrm{d}t}$$

Strain rate is an important variable in the forming process, because the strength and microstructural response of many metals are dependent on its value. Similar to strain, this variable can be expressed better in a second-order tensor.

3.2.5   Tribology and Micro-Tribology

The main parameter of interest in bulk forming is friction, which is dependent on surface load, lubrication, and surface characteristics, including surface roughness. Amorton's law dictates a linear dependence between load and friction force:

$$F_f = \mu \cdot L$$

where μ is the friction coefficient and L is the normal load.

In bulk forming, the friction coefficient is often established on the basis of experiments, and there is seldom an explicit formulation for this quantity. The double cup extrusion (DCE) test is a recognized way for establishing the friction coefficient experimentally, which is based on a double cup pressed between two punches with equal geometry. In the case of zero friction, the height ratio between the upper and lower cup will be unity, while no lower cup will be formed in the case of infinite friction.

Micro-tribology, also known as nano-tribology or molecular tribology, studies the behavior and damage of the friction interface at the molecular scale. The interest in this area is the interaction that exists between microsize surfaces, where the relative surface roughness of a micro-component will be greater compared with that of a macro-component.

A number of microscaled DCE tests were carried out by Engle, finding that measured friction coefficient depends on the scale of the experiment; the coefficient increased by a factor of 20 when the experiment was scaled by a factor of 8. According to Engel, this is due to the fact that, for microscale surface, more surface asperities reside close to the boundaries of the workpiece where they are less likely to form lubricant pockets under hydrostatic pressure, influencing the surface contact area and leading to an increase in the friction force. When working in the microsize domain, this increase in the coefficient contributes to the challenges in the handling and ejection of the bulk formed micro-components. It is essential to remember that surface area promotes friction and material volume brings strength; this is an important frictional challenge encountered within micro-bulk forming. With the increase in friction due to open lubricant pockets for the microsize domain, it is evident that friction is a key challenge to be overcome in micro-bulk forming [3].

3.3 MANUFACTURING PROCESS CHAIN

In order to develop a better understanding of the costs of medical implants, it is important to know how they are made. A short description of the manufacturing process chain, excluding the design, the obtaining of FDA approval, and testing, is given in the following sections. There are five basic steps involved in the manufacture of an orthopedic implant [8]:

1. manufacture of alloys and raw materials,
2. forming,
3. machining and finishing,
4. coating,
5. packaging and sterilization.

3.3.1 Manufacture of Alloys and Raw Materials

Bone represents the principal element for body sustention. Bone's tissue has several minerals and shows good mechanical properties and a spontaneous regeneration. So, bone defects and harms are healed by the formation of a new bone tissue with a structural organization similar to the original one. Both biological factors and prosthetic design features influence the performance of total joint prostheses: size and shape, materials, and surface characterization. The two primary issues in materials are mechanical properties (rigidity, corrosion characteristic, etc.) and biocompatibility.

Materials scientists have investigated metals, ceramics, polymers, and composites as principal biomaterials. Metals have been the primary materials in the past for this purpose due to their superior mechanical properties; in general, implants are made of cobalt-based alloys (Co-Cr-Mo and Co-Cr-Ni) or titanium alloys. These alloys are more resistant to corrosion, particularly the Co alloys that generate a durable chromium-oxide surface layer.

The most commonly employed ceramic materials are alumina, zirconia, and several porous ceramics. Ceramic material components suffer from early failures due to their low fracture toughness. Therefore, the material quality has been improved by modifying the production process and design requirements [9].

Materials such as silicone rubber, PE, acrylic resins, polyurethanes, polypropylene (PP), polymethylmethacrylate (PMMA), polyglycolide (PGA), polylactide (PLA), and polydioxanone (PDS) are part of the polymers mostly used in the manufacture of implants: the list leads with the ultrahigh-molecular-weight polyethylene (UHMWPE) and high-density polyethylene (HDE), given the strength and stiffness they exhibit [9].

3.3.2 Forming

The goal of these operations is to produce the intermediate forms for prosthesis, devices, and other metal components.

In forming, it is fundamental to design the forming process properly, involving the definition of the number of forming passes and the geometry of the dies. In fact, in many cases, it is not possible to reach the total material deformation in one single step, making it essential to divide into different passes involving the use of various dies.

The process is often carried out in a close die (Figure 3.1), and the die geometry itself must be carefully designed to guarantee the complete die cavity filling (considering the raw material volume variations due to the high working temperatures), the absence of internal defects, and adequate forming forces. Furthermore, it is important to reduce the material allowance to use the minimum raw material and energy necessary to successfully produce the part. The forming processes aiming at allowance reduction are called net shape processes, and today finite element analysis (FEA) can be considered a valid tool to optimize the die shape and the forming process in general.

3.3.3 Machining and Finishing

These operations are performed on the forged or cast forms. Components may be turned on a lathe to remove the seams generated on forged components; milling machines may cut odd shapes in the implants, which cannot be incorporated into the casting or forging operations. In addition, the surface quality can be improved by means of these material removal processes [8].

3.3.4 Coating

Porous coating, hydroxyapatite coating, ion implantation, and other processes are designed to improve the finish of the adherence of the implant to the bone.

3.3.5 Packaging and Sterilization

Implants are cleaned and inspected during each step of the process. They are then packaged and sterilized using gamma radiation according to the legal standards.

3.4 IMPLANTABLE DEVICES

An implantable device, or endoprosthesis, is an artificial structure or system in which the remaining functional parts of a previously fully developed physiological system are structurally supported or stimulated to restore (some) function. Prostheses are typically used to replace parts or restore functions lost by injury (traumatic), disease, or missing from birth (congenital).

The most often applied implants are

- orthopedic and traumatology prostheses (hip and knee joints, stabilizers of the bone fracture, screws, etc.),
- facial and maxillofacial devices,
- tooth implants, orthodontic apparatus, etc.
- surgical instruments
- cardiology devices as artificial heart valves, electrostimulators, etc.
- implants applied to ophthalmology,
- rehabilitation devices,
- drug delivery systems,
- medical instrumentation.

The majority of these devices are usually produced by forming operations, such as rolling, stamping, drawing, or punching. Heat and surface treatments are also very important during the technological process since the properties of implants depend on the applied surface treatment significantly.

It is important to underline that, apart from the previously mentioned implants, some surgical tools are also made by forming of metals.

3.5 BONE IMPLANTS

Among all the implantable devices, the bone implants, which are characterized by high mechanical properties, especially fatigue resistance, good corrosion resistance, and high biocompatibility level, are produced by forming of metals (mainly Ti alloys). They can be classified into four different categories:

- artificial joint replacement implants,
- spinal implants,
- cranial and maxillofacial implants,
- dental implants.

In bone implants, it is possible to classify two important categories of devices, the *external fracture fixation* devices and the replacement prosthesis.

3.5.1 External Fracture Fixation

External fixation systems comprise specially designed frames, clamps, rods, rod-to-rod couplings, pins, posts, fasteners, wire fixations, fixation bolts, washers, nuts, hinges, sockets, connecting bars, screws, and other components used in orthopedic and reconstructive surgery for the treatment of bone and joint injuries as well as the correction of skeletal deformities by attaching bones to an external device, which stabilizes the injured limb. Almost all of these devices are produced using cold or hot forming operations of metal alloys [10].

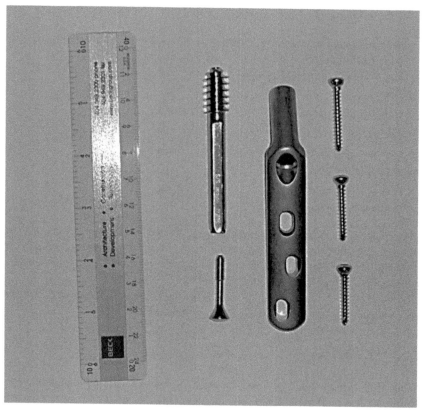

Figure 3.3 Examples of biomedical devices used in orthopedics. *Source*: Figure courtesy of Chris Martin.

The basic goal of fracture fixation is to stabilize the fractured bone, to enable fast healing of the injured bone, and to achieve early mobility and full function of the injured extremity. These components are made by sheet, wire, or bulk forming to guarantee that they fulfill the requirements in terms of mechanical and fatigue resistance and biocompatibility [11]. Figure 3.3 shows some examples.

3.5.2 Artificial Joint Replacement

The category of prostheses now affronted involves a lot of different kinds of implants. A quick list of the most important of them classified in three different groups is as follows [12]:

- In a *hinge joint*, the convex surface of one bone fits into the concave surface of another one, leading to a movement similar to the opening and closing of a hinged door, allowing only flexion and extension. Some examples of hinge joints are the elbow, knee, ankle, and joints between the fingers.

- The *ball-and-socket joint* allows twisting and turning movements and consists of one bone with rounded head (the ball) fitting into a cup-like depression of another bone (the socket). Some examples of these joints are the shoulder and the hip, with the shoulder joint being the most flexible joint in the entire body, allowing movement in any direction.
- There are *other types of joints* in the body. *Planar joints* allow two flat bones to slide over each other, for example, the bones of the foot and wrist. In a *pivot joint*, a rounded or pointed bone articulates with a ring formed by another bone and a ligament, such as in the radioulnar joint. A *condyloid joint* allows the head to nod and the fingers to bend. The thumb has a *saddle joint* that allows enough flexibility for the thumb to touch any other finger.

For each of these different joints, there is more than one kind of prosthesis. Considering the impossibility to analyze all of them, in this chapter only the most important ones are taken into consideration (Figure 3.4).

3.5.2.1 Elbow Joint Prosthesis   The elbow joint (Figure 3.5) is in the middle of the arm and is formed by three bones: the humerus of the upper arm, the paired radius, and ulna of the forearm [14].

Two main movements are possible at the elbow:

- Flexion and extension, caused by the hinge joint formed between the humerus and ulna.
- Turning the forearm over (pronation or supination), movement possible due to the articulation between the radius and the ulna.>

Elbow replacement involves surgical replacement of bones that make up the elbow joint with artificial elbow joint parts (prosthetic components) produced by hot forming processes in a close die. The artificial joint consists of two stems made of high-quality metal. They are joined together with a metal and plastic hinge that allows the artificial elbow joint to bend.

3.5.2.2 Knee Joint Prosthesis   The knee joint is a complex compound, a modified hinge joint that consists of three joints within a single synovial cavity: the patellofemoral joint (between the patella and the femur), and the medial and lateral tibiofemoral joints, which link the femur with the tibia [12].

The knee joint may be considered as a "mobile" joint, given that during rotation the femur and menisci move over the tibia, while the femur rolls and glides over the menisci during flexion and extension [15].

Although the overall designs for total knee implants vary, the typical total knee replacement implants have three basic components (Figure 3.6) [17]:

- *Femoral Component*: generally made of metal and curves around the end of the femur and manufactured by casting process.

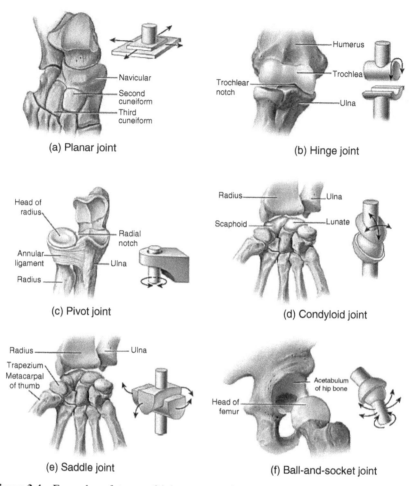

Figure 3.4 Examples of types of joints present in the human body. *Source*: Tortora and Derrickson 2009 [12]. Reproduced with permission of John Wiley & Sons. (*See color plate section for the color representation of this figure.*)

- *Tibial Component*: flat metal platform manufactured employing forging manufacturing process with a polyethylene spacer.
- *Patellar Implant*: dome-shaped piece of polyethylene that mimics the kneecap.

3.5.2.3 Ankle Joint Prosthesis The ankle joint is a hinge-type synovial joint, located between the distal ends of the tibia (shinbone) and the fibula (the small bone of the lower leg), and the superior part of the talus (the bone that fits into the socket formed by the tibia and fibula, which allows the foot to move up and down) [18].

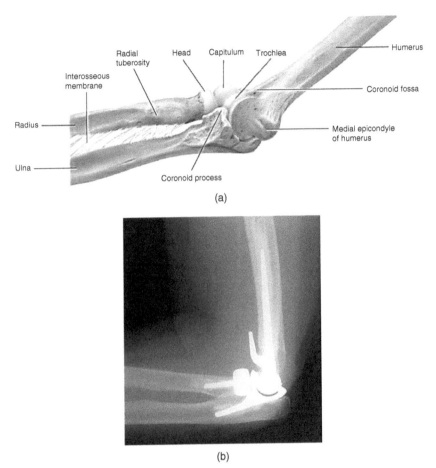

Figure 3.5 (a) Elbow joint scheme. *Source*: Tortora and Derrickson 2009 [12]. Reproduced with permission of John Wiley & Sons. (b) Elbow prosthesis. *Source*: Sanchez-Sotelo, 2011 [13]. Reproduced with permission of The Open Orthopaedics Journal.

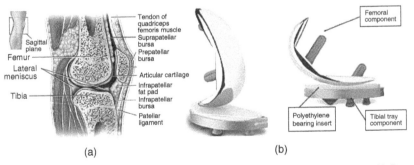

Figure 3.6 (a) Human knee joint scheme. *Source*: Tortora and Derrickson 2009 [12]. Reproduced with permission of John Wiley & Sons. (b) Knee prosthesis. *Source*: Culjat et al., 2012 [16]. Reproduced with permission of John Wiley & Sons. (*See color plate section for the color representation of this figure.*)

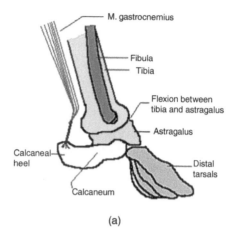

(a)

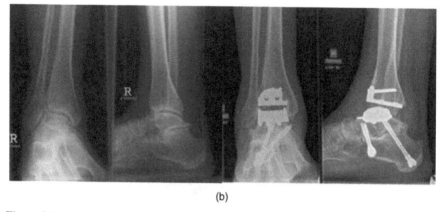

(b)

Figure 3.7 (a) Lateral view of human ankle. *Source*: Figure courtesy of Philip Chalmers. (b) Example of X-ray images of artificial joint inserted on ankle. *Source*: Barg et al., 2010 [19]. Reproduced with permission of Springer International Publishing. (*See color plate section for the color representation of this figure.*)

The artificial ankle prosthesis typically consists of three parts (Figure 3.7):

- The *tibial component* is a metal piece that replaces the socket portion of the ankle and is attached directly to the tibia bone.
- The *plastic cup or spacer* is a component that fits between the tibial and talus components.
- The *talus (talar) component* is a metal piece that replaces the top of the talus and fits into the socket of the tibial component.

Both the tibial and the talus components are manufactured by near-net-shape forging.

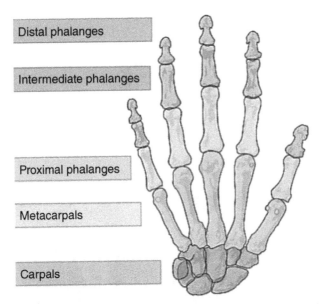

Distal phalanges

Intermediate phalanges

Proximal phalanges

Metacarpals

Carpals

Figure 3.8 Scheme of human hand bones. *Source*: Figure courtesy of Mariana Ruiz Villarreal.

3.5.2.4 Finger Joint Prosthesis The human hand has five digits (fingers) whose primary functions are manipulation and sensation. Each finger may flex and extend, abduct and adduct, and circumduct—movements possible by the presence of metacarpophalangeal joints, located between the distal heads of the metacarpals and the proximal phalanges of the digits (Figure 3.8) [20].

Prostheses are the treatment of choice when metacarpophalangeal and proximal interphalangeal joints are damaged, principally due to the presence of degenerative and rheumatoid arthritis. In some individuals, endoprosthetic joints do not permit appreciable adduction or abduction. Thus, there are some implants that provide an endoprosthesis for ginglymus joints composed of two parts: the first part is made from a metal insert to the human body and having a substantially cylindrical head (produced by forging manufacturing process), and the second part is made of molded plastic material having a socket of substantially cylindrical shape for receiving the head (Figure 3.9).

3.5.2.5 Shoulder Joint Prosthesis The shoulder is mainly composed of three different parts (Figure 3.10) [23]:

- The *scapula* is a large, flat triangular bone that functions mainly as a site of muscle attachment. On the head of the scapula is a depression called the "glenoid cavity," where the head of the humerus joins.
- The *clavicle* is a bone that connects the top of the sternum to the scapula, supporting the arm and transmitting force from it into the central skeleton.

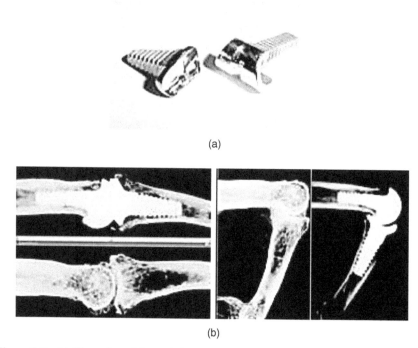

(a)

(b)

Figure 3.9 (a) Examples of finger joints; (b) example of X-ray images of artificial joint inserted on shoulder. *Source*: Gibson 2005 [21]. Reproduced with permission of John Wiley & Sons.

- The *humerus* is the bone of the upper arm and is the second most common long bone known. The smooth, dome-shaped head of the bone lies at an angle to the shaft and fits into a shallow socket of the scapula to form the shoulder joint.

Shoulder joint replacement surgery can either replace the entire acromioclavicular (AC) joint, in which case it is referred to as total shoulder joint replacement or total shoulder arthroplasty; or replace only the head of the humerus, a procedure called hemiarthroplasty.

The two artificial components that can be implanted in the shoulder during shoulder joint replacement surgery are as follows (Figure 3.10):

- *Humeral Component*: This part replaces the head of the humerus and is formed by a rounded ball attached to a stem that can be inserted into the bone. The material and the process that are employed to manufacture this component vary according to the quality of the patient's bone structure. If the bone structure is no longer very stable, the shoulder surface replacement is attached using bone cement, whereas if the bone is of good quality, the artificial shoulder joint can be secured using a cementless technique. The cemented stems are usually made of cobalt- or chromium-based alloys manufactured by casting,

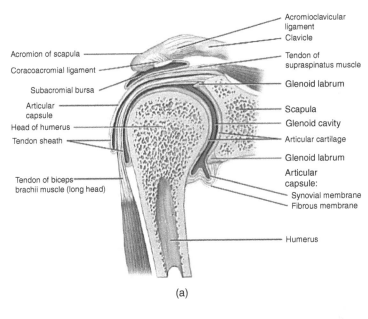

(a)

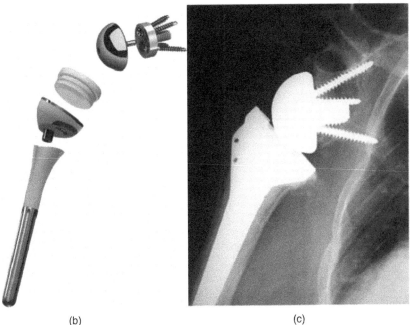

(b)   (c)

Figure 3.10 (a) Shoulder scheme. *Source*: Tortora and Derrickson 2009 [12]. Reproduced with permission of John Wiley & Sons. (b) Shoulder prosthesis. *Source*: Ekelund 2009 [22]. Reproduced with permission of Journal of Orthopaedic & Sports Physical Therapy. (c) Example of X-ray images of artificial joint inserted on shoulder. *Source*: Ekelund 2009 [22]. Reproduced with permission of Journal of Orthopaedic & Sports Physical Therapy. (*See color plate section for the color representation of this figure.*)

while the cementless stems are made of titanium alloys fabricated by forging manufacturing process.

- *Glenoid Component:* This part replaces the glenoid cavity, chiefly in most of the cases of UHMWPE.

3.5.2.6 Hip Joint Prosthesis The hip joint is formed by the head of the femur and the acetabulum of the hip bone, and its primary function is to support the weight of the body in both static (i.e., standing) and dynamic (i.e., walking or running) postures (Figure 3.11) [12].

The six movements that are performed by the hip joint are (listed in order of importance and with the range of motion from the neutral zero-degree position indicated): lateral or external rotation (30° with the hip extended, 50° with the hip flexed); medial or internal rotation (40°); extension or retroversion (20°); flexion or anteversion (140°); abduction (50° with hip extended, 80° with hip flexed); and adduction (30° with hip extended, 20° with hip flexed) [15].

The hip replacement is a surgical procedure in which the hip joint is replaced by a prosthetic implant, replacing both the acetabulum and the femoral head. The stem and the head of hip prosthesis are made of metal by hot forging process of titanium alloys. The anchorage cup is made of metal, and between prosthesis cup and head, there is a polyethylene spacer (Figure 3.11).

3.5.2.7 Wrist Joint Prosthesis The wrist is a difficult joint to replicate mechanically with eight inherently unstable small bones called carpals arranged in two rows: one in which the long thin bones of the hand radiate out, toward the fingers and thumb, and the second one in which the radius and ulna form a joint with it (Figure 3.12a).

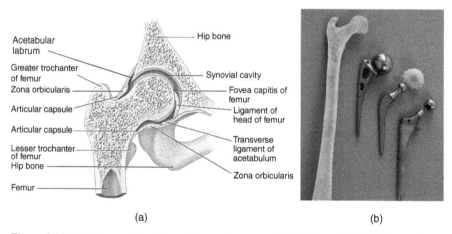

(a)   (b)

Figure 3.11 (a) Human hip scheme. *Source*: Tortora and Derrickson 2009 [12]. Reproduced with permission of John Wiley & Sons. (b) From left to right: Frontal plane cross-section of a femur; press fit implant; the original Chamley femoral stem and polyethylene cup; noncemented stem with a porous ingrowth surface. *Source*: Culjat et al., 2012 [16]. Reproduced with permission of John Wiley & Sons. (*See color plate section for the color representation of this figure.*)

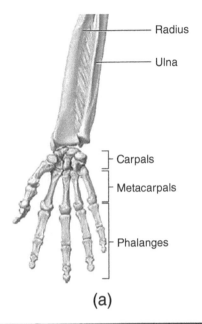

Radius

Ulna

Carpals

Metacarpals

Phalanges

(a)

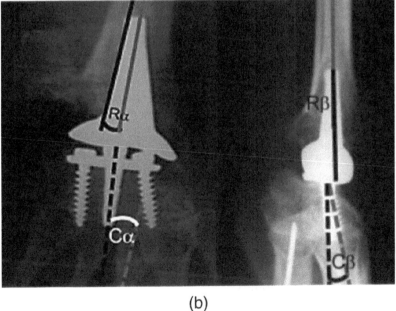

(b)

Figure 3.12 (a) Human wrist scheme. *Source*: Tortora and Derrickson 2009 [12]. Reproduced with permission of John Wiley & Sons. (b) X-ray images of wrist prosthesis positioning. *Source*: Facca et al., 2010 [24]. Reproduced with permission of John Wiley and Sons.

The primary reasons for wrist replacement surgery are to relieve pain and to maintain the function in the wrist and hand, conditions that are in most of the cases caused by arthritis, affecting the strength of the fingers and hand, making it difficult to grip or pinch. Wrist replacement surgery may enable the patient to retain or recover wrist movements and improve the ability to perform daily living activities, especially if the patient also has arthritis in the elbow and shoulder.

Wrist implant is made of the same kind of materials used for hip and knee joint replacements (cobalt-chrome or titanium alloys). There are several different designs, but most of them have two components and are made by hot forging of metal. The first piece is the one that attaches to the radius bone, the top of this component has a curve that matches with the wrist part, which attaches to the carpals and may have one long stem and one or two shorter stems that insert into the hand bones, or use small screws. A polyethylene component is used as spacer between the two components (Figure 3.12b).

3.5.3   Spinal Implants

In human anatomy, the vertebral column consists of 5 regions with a total of 33 vertebrae: cervical (7 vertebrae), thoracic (12 vertebrae), lumbar (5 vertebrae), sacral (5 vertebrae), and coccygeal (4 vertebrae). It is necessary to add intervertebral and spinal discs to these (Figure 3.13) [12].

The spinal structures function in unison to provide trunk flexibility, support of the upper body weight, and protection for the spinal cord and the roots of the spinal nerves, which transmit the body's electrical impulses from the brain to the rest of the body passing through the spinal canal and foramen.

Spinal components that are in abnormal state, whether of neoplasmic origin, trauma, or even age-related degeneration, can compromise the stability of the vertebral column and, therefore, the quality of life. To restore spinal stability, alignment, and function, it often becomes necessary to surgically fuse segments. In most of the cases, this procedure is accompanied by internal fixation, which typically includes rods, hooks, braided cables, plates, screws, and interbody cages of titanium, titanium alloys, or stainless steel, produced by forging manufacturing processes.

Interbody cages are structures that enhance the stabilization of segments (either between bones or in place of them) into which bone grafts can be packaged, allowing the growth of new bone through and around them (Figure 3.14) [26].

3.5.4   Craniomandibular Implants

The human skull has 22 bones attached to each other by sutures, making them immobile and forming the cranium. The cranium is the part of the skull that holds and protects the brain in a large cavity, called the cranial vault (Figure 3.15) [20].

The implants (prostheses) used in cranial surgeries to correct cranial defects resulting from trauma or disease are innumerable; for each of the bones that compounds the human skull a different kind of implant exists. Moreover, it is possible to build

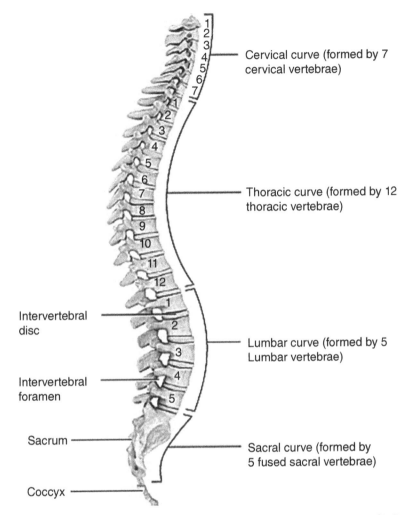

Cervical curve (formed by 7 cervical vertebrae)

Thoracic curve (formed by 12 thoracic vertebrae)

Intervertebral disc

Lumbar curve (formed by 5 Lumbar vertebrae)

Intervertebral foramen

Sacrum

Sacral curve (formed by 5 fused sacral vertebrae)

Coccyx

Figure 3.13 Vertebral column scheme. *Source*: Tortora and Derrickson 2009 [12]. Reproduced with permission of John Wiley & Sons.

a new kind of prosthesis with a new shape every time this is requested. In particular, the replacement of a cranial or maxillofacial bone can be produced in just a few hours, by using patient's computer tomography. CT allows reproducing the proper 3D geometry of the bone, and by means of CAD, it becomes easy to build the part using sheet stamping processes or rapid prototyping technologies such as incremental sheet forming (ISF). Besides cranial implants [28] [29], there are other works in the literature that used ISF as a technology to produce ankle prosthesis [30], a palate implant [31], a part of a knee implant [32], and most recently several maxillofacial implants [33]. An example of an overall reconstruction process of a cranial implant is reported in Figure 3.16.

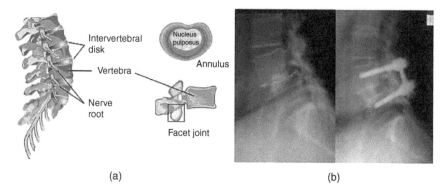

(a)   (b)

Figure 3.14 (a) Spine anatomy. *Source*: Kojić et al., 2008 [25]. Reproduced with permission of John Wiley & Sons. (b) Spinal vertebrae without spinal device and with spinal device. *Source*: Culjat et al., 2012 [16]. Reproduced with permission of John Wiley & Sons.

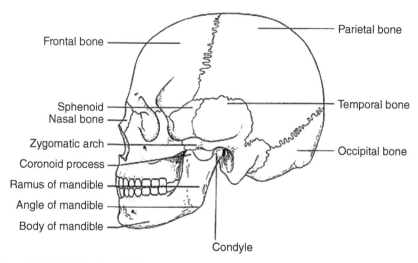

Figure 3.15 Skull and cranial bones. *Source*: Hollins 2012 [27]. Reproduced with permission of John Wiley & Sons.

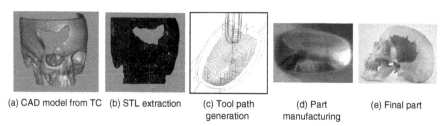

(a) CAD model from TC   (b) STL extraction   (c) Tool path   (d) Part   (e) Final part
                                                generation   manufacturing

Figure 3.16 Manufacturing process of cranial prosthesis. *Source*: Fiorentino et al., 2012 [34]. Reproduced with permission of John Wiley & Sons. (*See color plate section for the color representation of this figure.*)

Unlike other bones of the body, the irregular shape and geometrical complexity of most of the bones in the craniomandibular region are high, leading to structural and functional differences in that region.

The mandible is a bone similar in shape to a hoop and, therefore, multiple fracture sites are common. Besides fractures, diseases such as infections (osteomyelitis) and tumors, can affect the mandible zone. Depending on the type of injury affecting the zone, the treatment will be different; the surgical treatments used could be classified as conservatives (cystectomy, enucleation) or radicals (segmental or continuity resection).

The recurrence rate of conservative approaches is very high compared with radical surgery [35]. Nevertheless, if a radical surgery is applied, a mandibular prosthesis is required to reconstruct the affected zone. The prostheses are based mainly either on fitting reconstruction plates on the geometry of the patient's mandible during surgery practice (*in situ*) or the manufacture of customized plates or prosthesis prior to the operation. In the first situation, surgeons define the geometry of the mandibular prosthesis *in situ*. However, this is an arduous task given the complexity of the human's mandible geometry combined with complicated biomechanics [36], leading to an insufficient geometric definition, and consequently, to a prosthesis that does not fit correctly to the patient.

In the second situation, a customized prosthesis is usually manufactured based on the exact geometry of the patient's mandible obtained from the patient's computed tomography (CT) and magnetic resonance (MR); after this combination of image processing software, a design software (CAD, CAM, In-Vesalius, etc.) step is performed, concluding the manufacture of the prosthesis employing additive manufacturing technologies (AMT) such as fused deposition modeling (FDM), a process in which the model or part is produced by extruding small beads of material that harden immediately to form layers. The combination of technologies allow pre-bending the plate over the model, which has been manufactured by AMT; the works of Yamada et al. [37] and Wilde et al. [38] are recent examples (Figure 3.17).

The AMT model fabrication and the plate pre-bending are carried out before the operation, reducing a trial-and-error phase that occurs in *in situ* surgery. This sector also promotes a lot of research [39–43].

3.5.5 Dental Implants

The primary function of teeth is to prepare food for digestion by chewing. Mastication breaks down tough connective tissues and fibers and helps to saturate these materials with lubricants and enzymes.

The tooth can be classified into three different parts (Figure 3.18):

- *Crown* is the white part that is seen. It is covered by a layer of enamel that contains a crystalline form of calcium phosphates.
- *Neck* marks the boundary between the root and the crown.
- *Root* consists of a layer of dentin that extends up into the crown from the pulp cavity, which contains the root canal, nerves, and blood vessels.

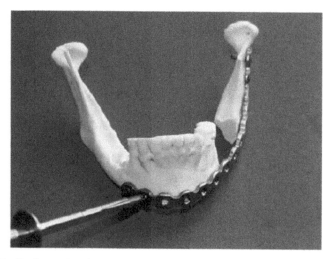

Figure 3.17 Pre-bent of a plate over AMT model. *Source*: Wilde et al., 2012 [38]. Reproduced with permission of Springer-Verlag. (*See color plate section for the color representation of this figure.*)

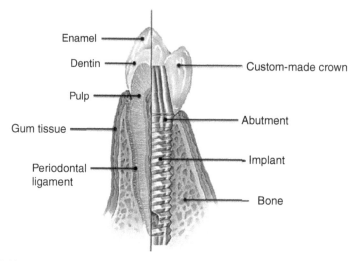

Enamel

Dentin

Pulp

Gum tissue

Periodontal ligament

Custom-made crown

Abutment

Implant

Bone

Figure 3.18 From left to right: dentition scheme; titanium screw dental implant (resembling a tooth root). *Source*: Wingrove 2013 [44]. Reproduced with permission of John Wiley & Sons. (*See color plate section for the color representation of this figure.*)

There are a lot of problems that can cause teeth diseases, and for each one, it is possible to identify the potential cause, the symptom, and the suggested treatment; the most common problems are plaque and cavities, which cause damage within the teeth structure resulting in tooth decay.

A dental implant is an artificial tooth root used in dentistry to support restorations that resemble a tooth or group of teeth. Nowadays, the most popular type of dental implant is the root-form endosseous implant, which is placed within the bone and typically consists of a titanium screw (resembling a tooth root) with a roughened or smooth surface. The majority of dental implants are made of commercially pure titanium, which is available in four grades depending on the amount of carbon and iron contained and are manufactured by the forging process (Figure 3.18) [45].

3.6 OTHER BIOMEDICAL APPLICATIONS

In the previous section, the implants that are mainly manufactured by the forming process were mentioned; however, there are some medical devices that are not completely manufactured by this process, although many of their components are made by forming, which are as follows:

- *Surgical Instruments*: Many surgical instruments include in their manufacturing process the forging step, for example, the arthroscopic shavers, which can be used to abrade, cut, and excise tissue and bone, remove loose fragments, and shave away debris in arthroscopic, jaw, or sinus surgeries. The manufacturing process starts with forging to shape metal parts through compressive forces. There are two types of forging processes involved in the manufacture of this kind of products: hand forging for small quantities and hammer forging for bulk quantities. After taking a crude impression of the shape, the excess material around the shape is trimmed for a more desirable shape. To create different features in the product, milling is used to cut away the unwanted material by machining. Grinding and filing are material-removing processes used in this manufacture chain as a step to produce the base for the general shape, using different types of gauges. Once that piece is ready, heat treatment is applied to change the physical and mechanical properties of the instrument without changing the desired shape and size obtained in previous manufacturing steps. After the heat treatment, the chemical treatment is utilized to clean the surface of the steel parts and increase corrosion resistance. At this stage, setting is done to align the instruments with proper functioning, and if a polishing process is needed to produce different types of appearances is applied, just before to check the desired specifications in the instrument and send them for final inspection and testing before packing.
- *Cardiology Devices as Artificial Heart Valves, Electrostimulators, etc.*: In the field of cardiology, the thermal forming process is used in the manufacture of leaflet heart valves for both ventricular assist devices and direct implantation in the body [46].
- *Ophthalmology Devices*: The forming manufacturing process is commonly used in the manufacture of instruments employed in ophthalmology surgeries, such as ophthalmic needles employed in instruments for removing cataracts. In this

product, a forming die is used to manufacture the needle assembly, which consists of forming the tapered portion of the outer tube [47].

- *Rehabilitation Devices*: Customized orthoses such as orthopedic boots and ankle supports, whose function is to restore normal alignment of the bone and ligaments parts and used for partial or total immobilization of a part that needs to recover its function, are manufactured by sheet forming (incremental) and polymers. This manufacturing technique is applied in this kind of product, because anthropometric differences between individuals require a bespoke product to improve their performance and usability [48].

- *Drug Delivery*: In localized drug delivery, where minimal tissue damage is desired, the forming process is used to manufacture thin metal foils, and their surface is deposited with drug-coated microparticles. Due to shock wave loading, the surface of the foil containing microparticles is accelerated to very high speeds, ejecting the deposited particles at hypersonic speeds, giving sufficient momentum to these to penetrate soft body tissues [49, 50].

- *Medical Instrumentation*: Many medical instrumentation systems include the forming process in their manufacturing process chain. One of the most representative examples is the basic forming operations for catheter tubing, which are tipping, bonding and laminating, necking, flaring, expanding, and forming. The process begins with the formation of a shape, usually cone or bullet, on the end of the catheter (tipping); then, two compatible plastic materials are thermally welded using controlled heat without melting and distorting the plastic catheter (bonding). After this, the outer diameter of the tube is reduced by pulling it through a heated reducing die (necking). A flange is formed on the end of a tube with a cone-shaped heated die (flaring); then, the end of the tube is expanded through a heated tapered die (expanding), and finally the tube is expanded usually under air pressure (forming). Two types of forming are free blowing (or heating a tube while applying controlled pressure into the tube) and clamping one end with a hemostat, similar to glass blowing [51].

REFERENCES

[1] Boljanovic V. *Sheet Metal Forming Processes and Die Design*. New York: Industrial Press Inc.; 2004. p xvii–xviii.

[2] Schuler GmbH. *Metal Forming Handbook*. Berlin, Heidelberg, New York: Springer; 1998. p 8–11.

[3] Qin Y. *Micromanufacturing Engineering and Technology*. Oxford: Elsevier Inc.; 2010. p 131.

[4] Qin Y. Micro-forming and miniature manufacturing systems-development needs and perspectives. J Mater Process Technol 2006;177:8–18.

[5] Langen HH, Masuzawa T, Fujino M. Modular Method for Microparts Machining and Assembly with Self-Alignment. Ann CIRP 1995;44:173–176.

[6] Lai X, Peng L, Hu P, Lan S, Ni J. Material behavior modeling in micro/meso-scale forming process with considering size/scale effects. Comput Mater Sci 2008;43:1003–1009.

[7] Wang GC, Zheng W, Wu T, Jiang H, Zhao GQ, Wei DB, Jiang ZY. A multi-region model for numerical simulation of micro bulk forming. J Mater Process Technol 2012;212:678–684.

[8] Mendenhall S. Orthop Network News 1992;3:12–13.

[9] Navarro M, Michiardi A, Cataño O, Planell JA. Biomaterials in orthopaedics. J R Soc Interface 2008;5:1137–1158.

[10] Solomin LN. *The Basic Principles of External Skeletal Fixation Using the Ilizarov Device.* Milan: Springer; 2008. p 1.

[11] Taljanovic M, Hunter TB, Miller MD, Sheppard JE. Gallery of Medical Devices. Part 1: Orthopedic Devices for the Extremities and Pelvis. RadioGraphics 2005;25:859–870.

[12] Tortora GJ, Derrickson B. *Principles of Anatomy and Physiology.* New Jersey: John Wiley and Sons, Inc.; 2009.

[13] Sanchez-Sotelo J. Total Elbow Arthroplasty. Open Orthop J 2011;5:115–123.

[14] Kapandji IA. *The Physiology of the Joints.* New York: Churchill Livingstone; 1982.

[15] Platzer W. *Color Atlas of Human Anatomy, Vol. 1: Locomotor System.* New York: Georg Thieme Verlag; 2009.

[16] Culjat M, Singh R, Lee H. *Medical Devices: Surgical and Image-Guided Technologies.* New Jersey: John Wiley and Sons, Inc.; 2012.

[17] Charlton P. *The Application of Zeeko Polishing Technology to Freeform Femoral Knee Replacement Component Manufacture.* Doctoral Thesis. University of Huddersfield June 2011.

[18] Moore KL, Agur AMR, Dalley AF. *Essential Clinical Anatomy.* Lippincott Williams & Wilkins; 2010. p 395.

[19] Barg A, Elsner A, Hefti D, Hintermann B. Haemophilic arthropathy of the ankle treated by total ankle replacement: a case series. Official J World Fed Hemophilia 2010;16:674–655.

[20] Drake RL, Vogl AW, Mitchell AWM. *Gray's Anatomy for Students.* Philadelphia: Elsevier Health Sciences; 2009.

[21] Gibson I. *Advance Manufacturing Technology for Medical Applications: Reverse Engineering, Software Conversion and Rapid Prototyping.* Chichester: John Wiley and Sons, Inc.; 2005. p 129.

[22] Ekelund A. Reverse shoulder arthroplasty. Official J Br Elbow Shoulder Soc 2009;1(2):68–75.

[23] Rockwood CA, Matsen FA. *The Shoulder.* Philadelphia: Elsevier Health Sciences; 2009.

[24] Facca S, Gherissi A, Livernaux PA. Contribution of computer-assisted surgery in total wrist prosthesis: a comparative preliminary study of eight cases. Int J Med Robotics Comput Assist Surg 2010;6:136–141.

[25] Kojić M, Filipović N, Stojanović B, Kojić N. *Computer Modeling in Bioengineering: Theoretical Background, Examples and Software.* Chichester: John Wiley and Sons, Ltd.; 2008. p 315.

[26] Martz EO, Goel VK, Pope MH, Park JB. Materials and design of spinal implants-A review. J Biomed Mater Res 1997;38(3):267–288.

[27] Hollins C. *Basic Guide to Anatomy and Physiology for Dental Care Professionals.* Chichester: John Wiley and Sons, Ltd.; 2012. p 108.

[28] Duflou JR, Lauwers B, Verbert J, Gelaude F, Tunckol Y. *Virtual Modelling and Rapid Manufacturing: Advanced Research in Virtual and Rapid Prototyping.* London: Taylor & Francis/Balkema; 2005. p 161–166.

[29] Gottmann A, Korinth M, Taleb Araghi B, Bambach M, Hirt, G, Manufacturing of cranial implants using incremental sheet metal forming, Proceedings of the 1st International Conference on Design and Processes for Medical Devices (PROMED), Brescia; 2012 May 2–4; SRL, Brescia: Neos Edizioni, SRL, Brescia, pp. 287–290.

[30] Ambrogio G, Denapoli L, Filice L, Gagliardi F, Muzzupappa M. Application of Incremental Forming Process for high customized medical product manufacturing. J Mater Process Technol 2005; 156–162.

[31] Tanaka S, Nakamura T, Hayakawa K, Nakamura H, Motomura K. Residual stress in sheet metal parts made by incremental forming process. AIP Conf Proc 2007;908(1):775–780.

[32] Oleksik V, Pascu A, Deac C, Fleaca R, Roman M, Bologa O. The influence of geometrical parameters on the incremental forming process for knee implants analyzed by numerical simulation. AIP Conf Proc 2010;1252:1208–1215.

[33] Duflou JR, Behera AK, Vanhove H, Bertol LS. Manufacture of accurate titanium cranio-facial implants with high forming angle using single point incremental forming. Key Eng Mater 2013;549:223–230.

[34] Fiorentino A, Marzi R, Ceretti E. Preliminary results on Ti incremental sheet forming (ISF) of biomedical devices: Biocompatibility, surface finishing and treatment. Int J Mechatron Manuf Syst 2012;5:36–45.

[35] Reichart PA, Philipsen HP, Sonner S. Ameloblastoma: biological profile of 3677 cases. Eur J Cancer, Part B 1995;31(2):86–99.

[36] Piloto PAG, Ribeiro JE, Campos JCR, Correia A, Vaz MAP. Simulação numérica do comportamento de uma mandíbula humana durante actividade mastigatória, Proceedings of 7th Congresso Nacional de Mecânica Experimental; 2008 Jan 23–25; pp. 231–233.

[37] Yamada H, Ishihama K, Yasuda K, Hasumi-Nakayama Y, Okayama M, Yamada T, Furusawa K. Precontoured mandibular plate with three-dimensional model significantly shortened the mandibular. Asian J Oral Maxillofac Surg 2010;22(4):198–201.

[38] Wilde F, Plail M, Riese C, Schramm A, Winter K. Mandible reconstruction with patient-specific pre-bent reconstruction plates: comparison of a transfer key method to the standard method-results of an in vitro study. Int J Comput Assist Radiol Surg 2012;7(1):57–63.

[39] Gellrich NC, Suarez-Cunqueiro MM, Otero-Cepeda XL, Schön R, Schmelzeisen R, Gutwald R. Comparative study of locking plates in mandibular reconstruction after ablative tumor surgery: THORP versus UniLOCK system. J Oral Maxillofac Surg 2004;62(2):186–193.

[40] Katakura A, Shibahara T, Noma H, Yoshinari M. Material analysis of AO plate fracture cases. J Oral Maxillofac Surg 2004;62(3):348–352.

[41] Martola M, Lindqvist C, Ha H, Al-Sukhun J. Fracture of titanium plates used for mandibular reconstruction following ablative tumor surgery. J Biomed Mater Res, Part B 2007;80(2):345–352.

[42] Schupp W, Arzdorf M, Linke B, Gutwald R. Biomechanical testing of different osteosynthesis systems for segmental resection of the mandible. J Oral Maxillofac Surg 2007;65(5):924–930.

[43] Sauerbier S, Kuenz J, Hauptmann S, Hoogendijk CF, Liebehenschel N, Schön R, Schmelzeisen R, Gutwald R. Clinical aspects of a 2.0- mm locking plate system for mandibular fracture surgery. J Cranio Maxillofac Surg 2010;38(7):501–504.

[44] Wingrove SS. *Peri-Implant Therapy for the Dental Hygienist: Clinical Guide to Maintenance and Disease Complications*. Chichester: John Wiley and Sons, Inc.; 2013. p 30.

[45] Ahn MR, An KM, Choi JH, Sohn DS. Immediate loading with mini dental implants in the fully edentulous mandible. Implant Dent 2004;13:367–372.

[46] Leat ME, Fisher J. The influence of Manufacturing Methods on the Function and Performance of a Synthetic Leaflet Heart Valve. J Eng Med 1995;209:65–69.

[47] Eichenbaum DM, Martin G, Rehkopf P. (to Ocular Associates), US Patent 4,377,897 (Mar. 22, 1983).

[48] Schaeffer L, Castelan J, Gruber V, Daleffe A, Marcelino R. Development of customized products through the use of incremental sheet forming for medical orthopaedic applications, 3rd International Conference on Integrity, Reliability and Failure, Porto, Portugal, July 20–24, 2009, paper S0209_P0308; 2009.

[49] Nagaraja SR, Rakesh SG, Prasad JK, Barhai PK, Jagadeesh G. Investigations on micro-blast wave assisted metal foil forming for biomedical applications. Int J Mech Sci 2012;61:1–7.

[50] Menezes V, Takayama K, Ohki T, Gopalan J. Laser-ablation-assisted microparticle acceleration for drug delivery. Appl Phys Lett 2005;87:1–3.

[51] Kucklick TR. *The Medical Device R&D Handbook*. Boca Raton: CRC Press; 2006.

4

LASER PROCESSING APPLICATIONS

TUĞRUL ÖZEL

Department of Industrial and Systems Engineering, School of Engineering, Rutgers University, Piscataway, NJ, USA

JOAQUIM DE CIURANA GAY AND DANIEL TEIXIDOR EZPELETA

Department of Mechanical Engineering and Industrial Construction, University of Girona, Girona, Catalonia, Spain

LUIS CRIALES

Department of Industrial and Systems Engineering, School of Engineering, Rutgers University, Piscataway, NJ, USA

4.1 INTRODUCTION

The development and use of microfluidic devices has increased in many areas where the need for transporting liquids or gases through channels of small cross-sectional areas has arisen. In particular, there is a growing demand for microscale medical devices in biomedical applications of all kinds and ranging to a wide variety of disciplines, such as genetics, biodetection, heat transfer applications, and sensors. This is reflected in the results obtained by researchers in these fields in the last 10–15 years. However, these are not the only areas where there is a growing demand for microscale medical devices. Other areas of interest are lab-on-chip devices, microneedles, and corrosion-resistant implantable devices. There is also growing interest in biomaterials such as transparent polymers, biocompatible metallic materials, and ceramics for mainstream medical devices. The attractiveness of biomaterials

Biomedical Devices: Design, Prototyping, and Manufacturing, First Edition.
Edited by Tuğrul Özel, Paolo Jorge Bártolo, Elisabetta Ceretti, Joaquim De Ciurana Gay, Ciro Angel Rodriguez, and Jorge Vicente Lopes Da Silva.

in a variety of medical device applications is counteracted by the very high cost of processing such materials and fabricating cost-effective end products. The ability to utilize moderate-cost nanosecond pulsed laser technology in direct fabrication of such products could drive the material cost down and allow expanding the use of a wide variety of biomaterials.

Currently, microfluidic devices can be constructed using many different methods, such as lithography, hot embossing, etching, micromilling, and laser processing. All these processes have their positives and negatives, but it is important to point out that laser microprocessing produces little to no chemical contamination of the product, generates no mechanical stresses, and provides the possibility of mass production via parallel processing. The biggest advantage of laser micro-manufacturing is the speed of the fabrication process and the flexibility it provides in terms of adjusting to different designs. Research performed on laser processing of microchannels on polymeric materials has focused on the use of "ultrashort" (pico- or femtosecond) pulsed lasers, which are expensive and not reliable to implement directly. Although the results have been very promising, short-pulse lasers are very expensive and, therefore, not suitable for mass production operations (see Table 4.1). Therefore, nanosecond pulsed laser processing is considered as an attractive option for companies interested in moderate-cost microscale manufacturing.

The data from previous research, obtained from laser processing using both near-infrared (NIR) and ultraviolet (UV) wavelength lasers, indicate that nanosecond pulsed laser processing for microfluidics is possible and demonstrate that laser processing is achievable whenever a highly absorbing polymer substrate is used. Nanosecond pulsed laser processing of a variety of polymers, metals, and ceramics is possible for biomedical device applications. These processes include photochemical and photothermal ablation, laser microjoining, laser melting, and solidification to form structures with dissimilar biomaterials such as polymer-to-metal and polymer-to-ceramic.

TABLE 4.1 Comparison between Continuous Wave and Pulsed Lasers

Laser Type	Continuous Wave	Nanosecond Pulsed (ArF, Nd:YAG, XeCl)	Pico-/Femtosecond Pulsed (Ti:Sapphire)
Peak power (W)	Moderate	$0.02–45 \times 10^6$	$0.005–10 \times 10^{10}$
Wavelength (nm)	1064–10,000	193–1064	800–900
Cost ($)	Low to moderate	Moderate	Very high
Quality/precision	Low	Moderate	High

4.2 MICROSCALE MEDICAL DEVICE APPLICATIONS

As mentioned previously, the development and utilization of microscale medical devices has boomed in recent years. In this section, a brief summary of the different areas in which research on microscale medical devices has been performed is

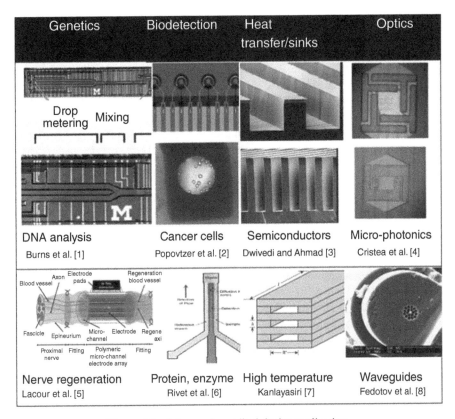

Figure 4.1 Microscale medical device applications.

presented. Figure 4.1 presents some of the individual fields in which microscale medical devices have been utilized: genetics, biodetection, heat transfer, and optics. In the field of genetics, researchers such as Burns et al. [1], Khanna et al. [9], and Chuncheng and Qianqian [10] have explored the use of microscale devices for DNA analysis, separation, and stretching. Lacour et al. [5] have worked on nerve regeneration, and Leclerc et al. [11] have studied the cultivation of liver and kidney cells. Weigl and Cabrera [12] investigated cell manipulation, with further research performed by Rivet et al. [6]. Some researchers Popovtzer et al., [2] focused on the detection of cancerous cells, while others such as Young [36] worked mainly on the detection of protein biotoxins. Liu et al. [13] investigated creatine kinase, an enzyme produced by various types of tissues and cells. Lastly, Ducrée [14] studied the use of microscale devices for bovine cell and fat screening. Among heat transfer applications, microscale devices have been used in semiconductor research [3], high temperature heat exchangers [7], regular heat exchangers [15], and most recently in external heat flux heat exchangers [16]. Finally, there are many types of microscale sensors: pressure sensors [17], temperature sensors [18], ethanol sensors [19], phenols sensors [20], and solution pH sensors [21].

4.3 PROCESSING METHODS FOR MEDICAL DEVICE FABRICATION

There are several manufacturing processes that could be utilized for prototyping and fabrication of the microscale medical devices described earlier: hot embossing, lithography, etching, ion beam machining, micromilling, and laser processing (some of them are compared in Table 4.2). Hot embossing is a preferred method for manufacturing thermoplastics. It utilizes a molding machine to heat the material above its softening temperature and then uses pressure to make the material take the mold's shape. Although the microstructure of the material can be replicated in detail, hot embossing has a limitation with respect to the aspect ratio of columns, walls, and grooves that can be achieved. In addition, a new mold is required to obtain a new geometry, so the process is highly inflexible. Lithography is a material removal process in which there is deposition of material in the substrate through a photoresist layer, followed by light exposure. Lithography is a complex process that requires considerable amounts of preparation and execution before obtaining the desired result. Wet etching is a process in which a masking agent is applied to the substrate, followed by the addition of a wet etchant so that the etch substance can remove the material from the main component via isotropy or anisotropy. A limitation of wet etching is that the etching rate depends principally on the direction of the material so that the etch depth will always be a function of the opening size rather than the etching time. Another considerable limitation is that etching conditions vary for different materials; therefore, the use of this process must be specifically tested for a given material. Ion beam machining, as described by Allen et al. [22], is performed by accelerating highly energetic ions in a stream of inert gas, and directing these ions toward the material to be removed. Ion beam machining can be classified into three groups according to the type of ions used: plasma, dynamic, and reactive ion beam machining. Among these, the use of plasma is the most popular worldwide, due to the fact that the material removal mechanism is chemical in nature and there is no damage to the subsurface of the material. Micromilling is a process by which mechanical force, by means of a tool, is utilized to remove the material directly. Since there is a strain force applied directly to the material, residual stresses remain in the workpiece, which affect its strength. An advantage of micromilling is that high rotational speeds improve the quality of the cut so that no postprocessing is needed. However, the main limitations to micromilling are the size of the tool itself and the need for even higher rotational speeds (around 1,000,000 rpm), which are not available. The chip cost can be quite high because of the need to constantly

TABLE 4.2 Manufacturing Process Comparison for Key Characteristics

Process	Laser	Hot Embossing	Lithography	Etching	Micromilling
Materials	All	Polymers	Photosensitive	Metals	All
Flexibility	High	Low	Low	Low	Moderate
Duration	Short	Long	Long	Moderate	Moderate
Cost	Moderate	Low	Moderate	Low	High

replace the tool due to wear and tear effects, while slow machining speeds are also an issue.

Laser micromachining-based medical device manufacturing techniques are based on laser ablation [23–25]. Laser ablation is defined as the process of removing material from a surface by irradiating it with a high energy laser beam. Laser ablation itself is the result of complex photochemical processes that are highly sensitive to the conditions in which the process is executed. Therefore, the outcome of the process itself depends on both the material properties and the laser characteristics (wavelength, pulse duration, beam size, etc.) [26–37].

The use of laser in the process for the development of microscale medical devices arises from the need for flexible and rapid manufacturing at a low-to-mid cost. All solid materials (metals, ceramics, and polymers) can be manufactured using a laser, given the appropriate process parameters. Laser processing techniques also benefit from not requiring access to clean-room facilities. Furthermore, they do not require additional elements such as masks (lithographic processes) or molds (hot embossing processes). As mentioned before, minor modifications to the design would require a completely new mask (in the case of lithography) or mold (for hot embossing); therefore, flexibility of the process is a huge advantage when it comes to utilizing laser micro-manufacturing. Laser processing provides an advantage over micromilling due to the faster manufacturing times of large batch sizes and the high costs associated with the milling tool. Therefore, laser processing can be viewed as a fast, flexible, cost-effective approach to classical micro-manufacturing processes. The flexibility of laser processing can be best seen in the wide array of geometries that can be attained through this manufacturing technique.

Complex microchannel systems can be developed (see Figure 4.2) allowing for better microfluidics applications. This complexity is required due to the need to integrate individual elements of different characteristics (fluids, biological/chemical, electrical, etc.) into a single medical device to increase the functionality of the

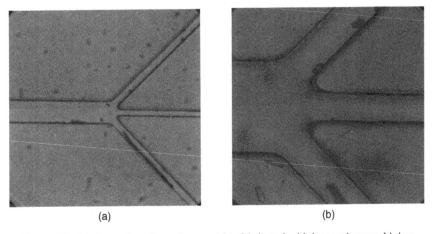

(a)   (b)

Figure 4.2   Various microchannel geometries fabricated with laser micromachining.

device. Therefore, research into microchannel generation using laser processing is of vital importance for the development of microscale medical devices.

The vast majority of the latest research performed on laser processing of microchannels on polymeric materials has focused on the use of "ultrashort" (pico- or femtosecond) pulsed lasers [24, 38, 39]. This is due to their higher cooling rate and lower heat-affected zone (HAZ), which is a direct result of the higher peak power produced relative to pulsed nanosecond or continuous wave lasers. A higher cooling rate is ideal because it generates lower peripheral thermal damage (due to the lower HAZ) and creates less debris (burrs) during the channel creation process. Although the results have been very promising, fast lasers are very expensive and, therefore, not suitable for mass production operations. On the other hand, nanosecond lasers provide a cheaper alternative to pico- or femtosecond pulsed lasers. Based on wavelength, lasers can be broadly classified into three categories: ultraviolet (UV) with a wavelength of 10–380 nm, visible light with a wavelength of 380–700 nm, and infrared (IR) with a wavelength of 700–1000 nm lasers. The vast majority of research has focused on the use of UV pico- or femtosecond pulsed lasers ([40, 41]; Zeng et al., [37]) as shown in Table 4.3, while only recently IR or NIR lasers have been used for laser micro-micromachining [45, 51, 52] with promising results. However, laser wavelength and pulse frequency are not the only parameters that affect the quality of the process. The size and shape of microchannels can be controlled by changing certain process parameters during the micro-manufacturing operation. Among the process parameters that determine the geometry of the microchannels, the most important are laser fluence and scanning speed, although laser power, Q-switch delay, pulse frequency, scanning rate, and focused beam distance from the surface are all relevant as well. The number of passes and the processing sequences are also important factors to take into account in microchannel creation, since one of the objectives is to obtain process repeatability—in other words, to be able to replicate the process and obtain consistent microchannel geometry in successive experiments. The main objective is to obtain process parameters that account for the highest material removal rate (MRR) and high quality, consistent microchannel geometries. In summary, the fact that most of the research has been conducted for pico- and femtosecond pulsed UV lasers, while the more cost-efficient nanosecond pulsed laser processing has not been fully explored (for both UV and IR lasers), presents a very attractive opportunity to carry out an important research. The development of more accurate theoretical mathematical models for thermal simulations would be very beneficial, since it would allow for a better understanding of the process and would simplify the process of obtaining the best laser process parameters. In order to achieve such a model, the properties of the laser utilized must be fully understood and characterized. Laser processing for medical device fabrication is a very promising field, yet one that still poses many questions that remain unanswered. For example, the disadvantages of laser processing in transparent polymers are peripheral thermal damage and the debris generated during the manufacturing process. The main objective of this research would be aimed at the development of nanosecond pulsed laser processing techniques that would attempt to circumvent the difficulties that investigators currently face when addressing laser processing for medical device fabrication.

TABLE 4.3 Process Parameters for Laser Processing of Polymers

Reference	Polymer	Laser Type	Wavelength (nm)	Pulse Length (s)	Fluence (J/mm²)
Aguilar (2005) [26]	PGA/PCL	ArF	193	3.5×10^{-8}	0.003–0.004
	PGA/PCL	Ti:Sapphire	800	2.2×10^{-13}	0.003–0.004
Baudach et al. [42]	PC/PMMA	Ti:Sapphire	800	1.5×10^{-13}	0.005–0.025
De Marco et al. [43]	PMMA	Ti:Sapphire	800	1.5×10^{-13}	0.86–2.6
Deepak Kallepalli and Soma [44]	PDMS/PMMA	Ti:Sapphire	800	1.0×10^{-13}	221.05
Gómez et al. [24]	PMMA/PI	Ti:Sapphire	800	9.0×10^{-14}	0.035
Krüger et al. [45]	PMMA	Ti:Sapphire	800	3.0×10^{-14}	0.0014–0.0045
	PMMA	Nd:YAG	1064	6.0×10^{-9}	0.0187–0.0384
Liu and Yi [46]	PMMA	KrF	248	3.0×10^{-14}	0.004–0.02
Pfleging et al. [47]	PMMA	Nd:YAG	355	4.0×10^{-7}	0.004–0.1
Pugmire et al. [48]	PMMA/PETG/PC/PVC	KrF	248	7.0×10^{-9}	0.0004–0.0041
Sowa and Tamaki [49]	PMMA	Ti:Sapphire	800	8.5×10^{-14}	–
Suriano et al. [39]	PMMA/PS/COP	Ti:Sapphire	800	4.0×10^{-14}	0.088–0.884
Suriyage et al. [50]	PC/PET	KrF	248	2.0×10^{-14}	–
Waddell and Kramer [41]	PMMA	KrF	248	7.0×10^{-9}	0.005–0.02
Wolynski et al. [51]	CFRP	Nd:YAG	1064	9.0×10^{-12}	0.009–0.8
			532	8.0×10^{-12}	0.005–0.09
			355	7.0×10^{-12}	0.009–0.4

4.4   BIOMATERIALS USED IN MEDICAL DEVICES

By definition, a biomaterial is any matter, surface, or construct that interacts with a biological system. Therefore, the most important property of a biomaterial is that it does not produce a negative reaction of the biological system when utilized. Biomaterials can be classified into three categories: inert (or nearly inert), active, and degradable. In general terms, metals are classified as inert biomaterials, ceramics can fall under any of the three groups, and polymers can be inert or degradable. Metals are utilized in medical devices such as orthopedic implants where strength and durability is required. Polymers have several advantages over other types of biomaterials. This is due to the fact that polymers have favorable thermal and chemical resistances, molding temperature, and other surface derivation properties. Transparent polymers are polymers whose degree of crystallinity approaches zero or one. They are generally the result of a polymer with an amorphous molecular chain, which provides them with an ample range of mechanical properties and phase behavior. Polymers are easily obtained at a low cost for a wide variety of options, allowing them to be the ideal material for mass production of disposable devices on a cost-effective basis. Polyurethane-based polymers are known as the most biocompatible polymers; therefore, they are often used in the production of artificial heart valves, blood vessels, and skin tissue. Polysiloxane is used in breast implants, artificial tendons, skin tissue, blood vessels, and heart valves [53]. Polyamide is utilized in retina implants, while polyethylene is used in disposable tubes, boxes, and syringes, and polyamide is the main material in the production of catheter tubes [54]. Polyether ether ketone (PEEK) scaffolding is used to stimulate bone growth due to its unique geometry and composition. In particular, research has focused on using polymethylmethacrylate (PMMA) as the main material for microchannel generation. PMMA has been a popular choice based on its biocompatibility, low cost, thermal stability, and mechanical properties. This would allow the creation of high quality microchannels. As mentioned previously, nanosecond pulsed laser microchannel generation in PMMA for microfluidics applications has been the recent focus of the Manufacturing Automation Research Laboratory (MARL) at Rutgers. Polydimethylsiloxane (PDMS) is the most widely used silicon-based polymer, another transparent polymer with promising applications in the biomedical devices field, since it is also biocompatible. PDMS microchannels are generally manufactured using soft lithography or hot embossing. Little research has been performed on nanosecond laser processing of PDMS microchannels, although some articles can be found in the literature on microchannel generation using ultrashort pulsed lasers [55, 56] (Figure 4.3).

4.5   MICROJOINING OF SIMILAR AND DISSIMILAR MATERIALS

Another required feature of microfluidic devices is the necessity to trap the fabricated elements so that the fluid may flow through the device. The most common approach to attaining this feature requires initially to produce the microchannel into the material, followed by the addition of a cover layer to seal the base layer. The bonding

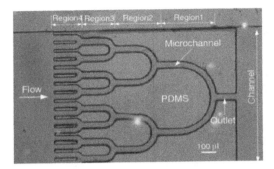

Figure 4.3   PDMS microchannel schematic.

of two layers at the microstructural level is a complex problem of its own, albeit one that can also benefit from laser processing technology. Laser microjoining of similar and dissimilar materials through either laser ablation or laser welding are viable techniques that have been investigated recently as well. The objective of laser welding of dissimilar materials is to provide a reliable process for the bonding of materials with different physical properties, especially at the microstructural level. Other methods of linking materials, such as adhesive bonding, anodic bonding, eutectic bonding, diffusion bonding, soldering, and ultrasonic welding, are not always available or fail to meet the required bond strength expectations because of the difference in basic material properties. Therefore, laser welding has been suggested as a possible alternative to join two or more dissimilar materials. In laser welding, the optical material properties play a vital role in the quality of the process. These properties include the degree to which light (and therefore, energy) can be absorbed, transmitted, or reflected. In general, the properties of the materials to be joined determine whether the laser energy is applied directly to the interlayer if one of the materials is transparent (transmission bonding) or if it is transferred through heat conduction (heat conduction bonding). In this case, it is necessary to have one material with high thermal conductivity, while the other material has a high level of heat resistance. In addition, a third material (interlayer material) can be chosen to facilitate the process. Glass-to-silicon microjoining with laser welding has been researched for overlapping joints of flat silicon and borosilicate glass samples [57]. Wild et al. utilized transmission bonding to achieve controlled heating at the interface, a process similar to anodic bonding (see Figure 4.4). During this process, both materials fail to reach their melting point. A drawback of this process is that the parameter window for acceptable bonds is very small, which affects the reliability of the process. Within the medical devices field, glass-to-silicon joining is required in many optoelectronic applications such as sensors, and also as micropumps for implantable drug delivery systems [58]. Ceramics-to-metal microjoining has been explored utilizing laser brazing by Lugscheider et al. [59] and Haferkamp et al. [60].

Polymer-to-metal microjoining has been explored for a wide variety of polymers, such as polyether block amide, thermoplastic polyurethane, and polyimide.

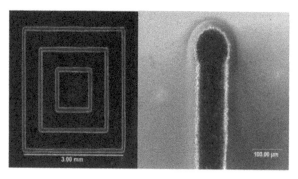

Figure 4.4   Silicon-to-glass bonding with Nd:YAG laser.

Researchers also determined that bond strength is affected by many laser parameters, including fluence, spot size, scanning speed, and laser wavelength. Finally, polymer-to-glass bonds have also been explored. To achieve proper bonding, a thin titanium film has been used as an interlayer material. The quality of the bonds is comparable to those achieved in polymer-to-metal bonding. However, repeatability of the process remains an issue, due to the small window with respect to laser parameters. The optimization of laser parameters through the development of better thermal models and simulations could lead to better results. The main difficulty faced when attempting laser welding of both similar and dissimilar materials is with the generation of a weld joint that does not meet the necessary standards for mechanical performance. Welds created with laser usually possess high hardness and are very brittle so they are not suitable for heavy-duty industrial applications. The high hardness and brittleness of the joint is a result of the formation of fragile intermetallic phases and layers during the welding process. Intermetallic compounds increase the residual stress in the HAZ, which makes the welded joint weaker. In theory, higher cooling rates would reduce the residual stress in the HAZ, which would in turn avoid the embrittlement of the weld. Current research points to the fact that cooling rate alone might not be enough to avoid the embrittlement of the weld, at which point interlayer materials are needed to neutralize the intermetallic phases/layers and gradually improve the joint strength [61]. However, it is possible that interlayer materials also contribute to the formation of fragile intermetallic layers [62]. The time required to execute the process also contributes to the growth of intermetallic phases. The longer the workpiece is subjected to the laser, the larger the HAZ will be, and the larger the residual stresses will be as well. Therefore, the time/temperature history of the process is a vital factor in the formation of intermetallic phases/layers. The geometry of the pieces welded together plays an important role in the properties of the welded joint. In summary, laser joining of dissimilar materials such as ceramics, polymers, and metals at the microscale can create small, narrow bonds that are strong and closely sealed, while minimizing heat input and heat-related damage to the workpiece. Therefore, they carry an advantage over more traditional bonding methods such as anodic and adhesive bonding. Results to date have shown that bonding of dissimilar materials at the microscopic level, by means of laser processing, is possible. However, issues regarding the integrity of the bond, especially when utilizing an interlayer material, remain to be addressed.

4.6 LASER MICROMACHINING FOR MICROFLUIDICS

There have been studies conducted to investigate the effectiveness of nanosecond pulsed laser micromachining technique for direct fabrication of microfluidic channels in transparent polymers. Modeling and experimental investigations have been conducted by Teixidor et al. [52, 63] and Criales et al. [64] on the effects of process parameters and the viability of directly fabricating microchannels in PMMA and PDMS polymers, which are suitable for the fabrication of microfluidic devices due to their biocompatibility and transparent properties. Microchannels fabricated on PMMA and PDMS with laser micromachining in NIR and UV wavelengths are shown in Figure 4.5.

Initially, the influence of laser process parameters (wavelength, pulse energy, pulse frequency, scanning rate, and spot size) on the geometry (width and depth) of microchannels fabricated in PMMA polymeric substrates and resultant MRR were also investigated by Teixidor et al. [52, 63]. In this study, investigations on laser micromachining of PMMA polymer by using NIR ($\lambda = 1064$ nm) and UV ($\lambda = 355$ nm) wavelengths in an Nd:YAG nanosecond pulsed laser system were conducted. PMMA polymer presents different absorption characteristics at those wavelengths, so the results were expected to vary accordingly. At UV wavelength, the PMMA polymer exhibits high absorption creating a strong photothermal effect,

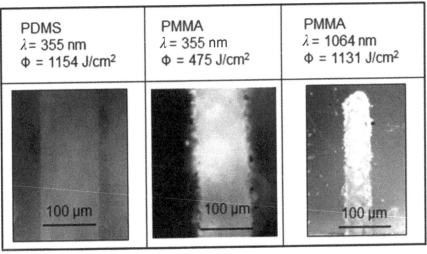

PDMS $\lambda = 355$ nm $\Phi = 1154$ J/cm^2	PMMA $\lambda = 355$ nm $\Phi = 475$ J/cm^2	PMMA $\lambda = 1064$ nm $\Phi = 1131$ J/cm^2

Figure 4.5 Microchannels fabricated on PMMA and PDMS with laser micromachining in NIR and UV wavelengths.

while UV light photon energy can exceed most intermolecular bond energies in the polymer and photochemical ablation can occur simultaneously even at low fluence levels. On the other hand, ablation of transparent polymers is directly related to the energy deposited in polymer surface per unit time. Therefore, a high fluence level combined with lower scanning speed can achieve ablation photothermally at NIR wavelength even though photon energy of NIR laser light is not enough to decompose intermolecular bond energies of the polymer. This study demonstrates the effectiveness of using NIR wavelength nanosecond pulse laser ablation on PMMA polymer with comparison to UV wavelength laser ablation in fabricating microchannels. In addition, the paper analyzes microchannel geometry (width and depth profile) and MRR during a single-pass laser micromachining process on PMMA polymer substrate as a function of laser processing parameters. Microchannel images and main effects plots are utilized to identify the effects of the process parameters to produce microchannels in PMMA polymer. Although the dimensions and geometric shape of the microchannels produced with the laser micromachining process exhibit variations, the results suggest that NIR nanosecond laser is capable of ablating PMMA polymer in order to produce microfluidic microchannels. It was also seen that wavelength and fluence are more important than pulse duration and pulse frequency to obtain photochemical ablation, avoid cracking, and avoid photothermal ablation-related discrepancies. Furthermore, depth and width are directly proportional to the scanning speed. However, this parameter has an opposite effect on the MRR. A comparison of NIR and UV wavelength laser ablation reveals that ablation fluence in UV irradiation is lower and more consistent microchannel geometries can be obtained with UV irradiation. Physical models for predicting the depth and the profile of laser-ablated channels validated with experimental results are found useful for process planning purposes. Modeling of the physical process, photothermal ablation, was also a scope of this project. A mathematical model of channel depth profile was also developed [63], similar to what Yuan and Das [65] proposed, albeit with a modification. From the energy balance equation, the channel depth can be solved by substituting the laser energy density, the decomposition energy, and the energy conducted into the surface of the PMMA. Therefore, the channel depth $D(y)$ can be obtained. In this physical model, a density of PMMA polymer of $\rho = 1185 \times 10^{-9}$ kg/mm^3, a latent heat of fusion of the material, $L = 355{,}640$ J/kg, a decomposition temperature of $T_v = 733$ K for PMMA polymer, and a specific heat of $c_p = 1466$ J/kg/K are used. The laser beam radius at the focal waist was set at 95 μm for the IR wavelength and 37.5 μm for the UV wavelength irradiation. The depth profile with respect to channel distance parameter was computed in a Matlab® software. Comparison of predicted and actual depth profile is also found to be reasonably acceptable, indicating the validity of this process model for predicting depth profile from given laser process parameters for process planning and optimization purposes.

Criales et al. [64] conducted experimental work using a solid-state Nd:YAG laser with 355 nm ultraviolet (UV) wavelength and 5 ns pulse duration at various energy densities and pulse overlapping (PO). The study was focused on understanding the effects of two main process parameters: fluence and PO. This study has investigated the effect of varying process parameters on the ablation depth and profile achieved and the resultant microchannel dimensional quality. It presents findings indicating that both process parameters have strong effects on the profile shape and variability of the microchannel width and 15 depth. For PMMA polymer, the lowest dimensional variability for the microchannel profile is obtained with low fluence values and highest PO factor, whereas for PDMS polymer, it was observed that the microchannel width and depth decreased linearly with increasing fluence and increased nonlinearly with increasing scanning rate. Furthermore, process modeling is utilized for predicting microchannel profile and ablation depth, and these predictions were validated with experimental results obtained with pulsed laser micromachining at UV wavelength.

Figure 4.6 shows experimental channel profiles, which have been irradiated with UV wavelength on PMMA and PDMS, and the superimposed modeling results for different combinations of laser power and scanning speed.

In summary, these research studies showed the effects of nanosecond laser processing parameters on depth and width of microchannels fabricated from PMMA and PDMS polymers. Microchannel width and depth profiles were measured, and main effects plots were obtained to identify the effects of process parameters on channel geometry (width and depth) and MRR. The relationship between process variables (width and depth of laser-ablated microchannels) and process parameters was investigated. It was observed that laser processing at UV wavelength provided more consistent channel profiles at lower fluence due to higher laser absorption of PMMA and PDMS at this wavelength. Further mathematical modeling for predicting microchannel profile should be developed and validated with experimental results obtained with pulsed laser micromachining at NIR and UV wavelengths, in order to optimize microchannel geometry quality.

Microchannels are a key element for microneedle array patches, a new medical device for painless drug delivery, self-diagnosis, and fluid extraction applications that has been investigated previously at the MARL at Rutgers. The basic premise of painless microneedle patches is to design microneedles so small in size that they do not interfere with the pain receptors located in the dermis of the skin [66]. Microneedles can be fabricated using a wide variety of biomaterials, including metals, silicon compounds, polymers, and glass, and through several micro-manufacturing techniques. Since the needles must be hollow to transmit microfluidic quantities, microchannels that would accomplish this task could be manufactured using nanosecond pulsed laser processing techniques [66].

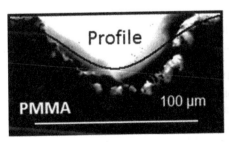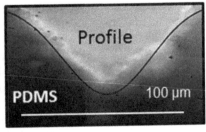

Figure 4.6   Laser micromachining based PMMA and PDMS microchannel profiles.

4.7   LASER MICROMACHINING FOR METALLIC CORONARY STENTS

Another laser micromachining medical device application is the manufacturing of coronary stents. A stent is a wire mesh tube, which is deployed in a diseased coronary artery to provide smooth blood circulation. Stents can be either balloon expandable or self-expanding (using shape memory alloys). Stents are typically made from biocompatible materials such as stainless steel, nitinol (Ni-Ti alloy), cobalt-chromium, titanium, tantalum alloys, platinum iridium alloys, as well as polymers. The most commonly used is stainless steel. The laser key requirement for its fabrication is a small consistent kerf width and this demands constant beam quality and excellent laser power stability. The laser cut must have a good surface quality with a minimum amount of slag and burr to reduce postprocessing; similarly, the HAZ and molten material recast need to be small [67].

There are several research works, which use a fiber laser because it is seen as an efficient, reliable, and compact solution for micromachining stents due to the combination of high beam power with high beam quality, small spot sizes, higher efficiency, and almost free maintenance. Many of these authors study how the process parameters of the laser cutting affect the quality of the resultant surfaces. Muhammad et al. [68] investigated the basic characteristics of fiber laser cutting of stainless steel 316L tube and understood the effect of introducing water flow in the tubes on minimizing back wall damages and thermal effect. The influence of laser parameters on the cutting quality for fixed gas type and gas pressure was investigated. Wet cutting enabled significant improvement in cutting quality. It resulted in narrower kerf width, lower surface roughness, less dross, absence of back wall damages, and smaller HAZ. Laser average power and pulse width play a significant role in controlling the cutting quality. Increasing the pulse width increased beam/material interaction time, which increased the kerf width and surface roughness. Kleine et al. [67] presented micro-cutting results in stainless steel samples of 100 and 150 μm where the kerf width and the surface quality were analyzed. They also studied the laser conditions to minimize HAZ. They concluded that the fiber laser is capable of achieving very small diameters and small kerf widths presenting very similar features to those produced with a Nd:YAG laser. Baumeister et al. [69] presented laser micro-cutting results for stainless steel foils with the aid of a 100 W fiber laser. Different material thicknesses were evaluated (100–300 μm). Processing was carried out with continuous wave operation, and with nitrogen and oxygen as assisting gases. Besides the high processing rate of oxygen-assisted cutting, a better cutting performance in terms of a lower kerf width was obtained. Minimal kerf width of less 20 μm was obtained with oxygen as the assisting gas. The kerf widths obtained with nitrogen-assisted cutting were generally wider. Scintilla et al. [70] presented results of Ytterbium fiber laser cutting of Ti6Al4V sheets (1 mm thick) performed with argon as the cutting assistance gas. The effect of cutting speed and shear gas pressure on the HAZ thickness, squareness, roughness, and dross attachment was investigated. The results showed that increasing the cutting speed and then decreasing the heat input from at 2 kW lead to an increase of HAZ and RL thickness occurs, up to 117 μm. Powell et al. [71] developed an experimental and theoretical investigation into the phenomenon of "striation free cutting," which is a feature of fiber laser/oxygen cutting of thin section mild steel. This paper concludes that the creation of very low roughness edges is related to an optimization of the cut front geometry when the cut front is inclined at angles close to the Brewster angle for the laser-material combination. Teixidor et al. [72] presented an experimental study of fiber laser cutting of 316L stainless steel thin sheets. The effect of peak pulse power, pulse frequency, and cutting speed on the cutting quality for fixed gas type and gas pressure was investigated. The analysis showed that increasing the peak pulse power and the cutting speed increases the kerf width, surface roughness, and dross deposition. Higher pulse frequency values result in bigger kerf and dross but improve the surface roughness. A mathematical model for the dross dimensions was formulated. The dross height and the dross diameter were analyzed and compared with the experimental results. Both dimensions increase with the increasing pulse peak power. Other authors also studied the formation of the dross, developing analytical models in order to predict the shape of this melt material attached to the cutting edge [73–75].

Other researchers studied the effect of process parameters on the fabrication of stents using different lasers on several materials such as nitinol or stainless steel. Kathuria [76] described the precision fabrication of metallic stent from stainless steel (SS 316L) by using short-pulse Nd:YAG laser. They concluded that the processing of stent with desired taper and quality shall still be preferred by the short-pulse and higher pulse repetition rate of the laser, which is desired to reduce further the HAZ as well as the wave depth of the cut section. Pfeifer et al. [77] examined the pulsed Nd:YAG laser cutting of 1 mm thick NiTi shape memory alloys for medical applications (SMA-implants). They studied the influence of pulse energy, pulse width, and spot overlap on the cut geometry, roughness, and HAZ. They generated small kerf width ($k = 150$–$300\,\mu m$) in connection with a small angle of taper ($\theta < 2°$). Compared with short- and ultrashort-laser processing of SMA, high cutting speeds ($v = 2$–$12\,mm/s$) along with a sufficient cut quality ($R_z = 10$–$30\,\mu m$) were achieved. The drawbacks can be seen in the higher thermal impact of the laser-material processing on the SMA, resulting in a HAZ (dimension: 6–$30\,\mu m$), which affects the material properties and reduces the accuracy of the cutting process. Shanjin and Yang [78] presented Nd:YAG pulsed laser cutting of titanium alloy sheet to investigate the influences of different laser cutting parameters on the surface quality factors such as HAZ, surface morphology, and corrosion resistance. The results presented show that medium pulse energy, high pulse rate, high cutting speed, and argon gas at high pressure help to acquire thin HAZ layers. Also in comparison with air- and nitrogen-assisted laser cutting, argon-assisted laser cutting comes with unaffected surface quality. Yung et al. [79] performed a qualitative theoretical analysis and experimental investigations of the process parameters on the kerf profile and cutting quality. They micro-cut thin NiTi sheets with a thickness of $350\,\mu m$ using a 355 nm Nd:YAG laser. The results showed that the kerf profile and cutting quality are significantly influenced by the process parameters, such as the single pulse energy, scan speed, frequency, pass number, and beam offset, with the single pulse energy and pass number having the most significant effects. They obtained debris-free kerf with narrow width ($\approx$25 µm) and small taper ($\approx$1°). They concluded that as the single pulse energy is increased and the laser beam velocity is decreased, the kerf width increases. Muhammad et al. [80] studied the capability of picosecond laser micromachining of nitinol and platinum-iridium alloy in improving the cut quality. Process parameters used in the cutting process have achieved dross-free cut and minimum extent of HAZ. Huang et al. [81] studied the effect of a femtosecond laser machining on the surface and characteristics of nitinol. The results have produced surface roughness of about 0.2 µm on nitinol. SEM and microstructural analyses revealed a HAZ smaller than 70 µm in depth, and a re-deposited layer of about 7 µm exists on the machined surface. Finally, Scintilla and Tricarico [82] analyzed the influence of processing parameters and laser source type on cutting edge quality of AZ31 magnesium alloy sheets and studied the differences in cutting efficiency between fiber and CO_2 lasers. They investigated the effect of processing parameters in a laser cutting of 1 and 3.3 mm thick sheets on the cutting quality. Their results showed that productivity, process efficiency, and cutting edge quality obtained using fiber lasers outperform CO_2 laser performances.

REFERENCES

[1] Burns M, Brahmasandra S, Handique K, Webster J, Krishnan M, Sammarco T, Man P, Jones D, Heldsinger D, Mastrangelo C, Burke D. An integrated nanoliter DNA analysis device. Science AAAS 1998;282:484–487.

[2] Popovtzer R, Neufeld T, Popovtzer A, Rivkin I, Margalit R, Engel D, Nudelman A, Rephaeli A, Rishpon J, Shacham-Diamand Y. Electrochemical lab on a chip for high-throughput analysis of anticancer drugs efficiency. Nanomed Nanotechnol Biol Med 2008;4:121–126.

[3] Dwivedi VK, Ahmad S. Fabrication of very smooth walls and bottoms of silicon microchannels for heat dissipation of semiconductor devices. Microelectronics 2000;31:405–410.

[4] Cristea D, Purica M, Manea E, Avramescu V. Experiments for microphotonic components fabrication using Si ⟨1 1 1⟩ etching techniques. Sens Actuators, A 2002;99(1):92.

[5] Lacour SP, Lago N, Tarte E, McMahon S, Fawcet J. Long micro-channel electrode arrays: a novel type of regenerative peripheral nerve interface. IEEE Trans Neural Syst Rehabil Eng 2009;17:454–460.

[6] Rivet C, Hirsch A, Hamilton S, Lu H. Microfluidics for medical diagnostics and biosensors. Chem Eng Sci 2010;66:1490–1507.

[7] Kanlayasiri K. A nickel aluminide microchannel array heat exchanger for high-temperature applications. J Manuf Processes 2004;6:72–80.

[8] Fedotov VA, Mladyonov PL, Prosvirnin SL, Zheludev NI. Planar electromagnetic metamaterial with a fish scale structure. Phys Rev E 2005;72(056613):1–4.

[9] Khanna P, Kim S, Seto E, Jaroszseki M. Use of nanocrystalline diamond for microfluidic lab-on-a-chip. Diamond Relat Mater 2006;15:2073–2077.

[10] Chuncheng Z, Qianqian C. Electrophoretical stretching of DNA in a hybrid microchannel. Polymer 2009;50:5326–5332.

[11] Leclerc E, Griscom L, Baudoin R, Legallais C. Guidance of liver and kidney organotypic cultures inside rectangular silicone microchannels. Biomaterials 2006;27:4109–4119.

[12] Weigl BH, Cabrera CR. Lab-on-a-chip for drug development. Adv Drug Delivery Rev 2003;55:349–377.

[13] Liu CX, Wang H, Guo ZH, Cai XX. A novel disposable amperometric biosensor based on trienzyme electrode for the determination of total creatine kinase. Sens Actuators, B 2007;122:295–300.

[14] Ducrée J. Next-generation microfluidic lab-on-a-chip platforms for point-of-care diagnostics and systems biology. Procedia Chem 2009;1:517–520.

[15] Tang Y, Pan M, Wang J. Ring-shaped microchannel heat exchanger based on turning process. Exp Therm Fluid Sci 2010;34:1398–1402.

[16] Mathew B. Experimental investigation of thermal model of parallel flow microchannel heat exchangers subjected to external heat flux. Int J Heat Mass Transfer 2012;55:2193–2199.

[17] Adam B, Henn R, Reiss S, Lang M. A new micromechanical pressure sensor for automotive airbag applications. In: Valldorf J, Gessner W, editors. Advanced Microsystems for Automotive Applications 2008: Components and Generic Sensor Technologies. Springer; 2008. p 259–284.

[18] Xue Z. Integrating micromachined fast response temperature sensor array in a glass microchannel. Sens Actuators, A 2005;122:189–195.

[19] Shi J, Wang F, Yang P, Wang L, Wang Q, Chu P. Pd/Ni/Si-microchannel-plate-based amperometric sensor for ethanol. Electrochim Acta 2011;56:4197–4202.

[20] Popovtzer R, Ronb EZ, Rishpon J, Shacham-Diamand T. Electrochemical detection of biological reactions using a novel nano-bio-chip array. Sens Actuators, B 2006;119:664–672.

[21] Zhan W, Crooks RM. Hydrogel-based microreactors as a functional component of microfluidic systems. Anal Chem 2002;74:4647–4652.

[22] Allen DM, Evans RW, Fanara C, O'Brien W, Marson S, O'Neill W. Ion beam, focused ion beam, and plasma discharge machining. CIRP Ann Manuf Technol 2009;58:647–662.

[23] Chen K, Yao YL. Process optimization in pulsed laser micromachining with applications in medical device manufacturing. Int J Adv Manuf Technol 2000;16:243–249.

[24] Gómez D, Lizualn I, Ozalta M. Femtosecond laser ablation for microfluidics. Opt Eng 2005;44:051105-1–051105-8.

[25] Ramanathan D, Molian P. Ultrafast laser micromachining of latex for balloon angioplasty. J Med Devices 2010;4:014501-1–014501-3.

[26] Aguilar CA, Lu Y, Mano S, Chen S. Direct micro-patterning of biodegradable polymers using ultraviolet and femtosecond lasers. Biomaterials 2005;26:7642–7649.

[27] Chen S, Kancharla VV, Lu Y. Laser-based microscale patterning of biodegradable polymers for biomedical applications. Int J Mater Prod Technol 2003;18:457–468.

[28] Hu W, Shin YC, King GB. Micromachining of metals, alloys and ceramics by picosecond laser ablation. J Manuf Sci Eng 2010;132(1):011009.

[29] Lugscheider E, Burschke I. On the way to the 21st century-new joining processes for the microsystem technology. In: High Temperature Brazing and Diffusion Welding. DVS-Berichte; 1998. p 192.

[30] Pérez-Murano F. Concepts and principles of optical lithography [Online]. Available: http://www.unizar.es/nanolito/pdf/ponencias_jaca/Concepts_and_principles.pdf [Accessed], 2015.

[31] Prausnitz M, Mikszta J, Cormier M, Andrianov A. Microneedle-based vaccines. Curr Top Microbiol Immunol 2009;333:369–393.

[32] Waqar A, Jackson MJ. *Emerging Nanotechnologies for Manufacturing, Micro and Nano Technologies*. Oxford, UK: William Andrew; 2009.

[33] Wu B, Shin YC. A simplified model for high fluence ultra-short pulsed laser ablation of semiconductors and dielectrics. Appl Surf Sci 2009;255(9):4996–5002.

[34] Wu B, Shin YC, Pakhal H, Laurendreau NM, Lucht RP. Modeling and experimental verification of plasmas induced by high-power nanosecond laser-aluminum interactions in air. Phys Rev E 2007;76:026405.

[35] Wu J, Sansen W. The glucose integratable in the microchannel. Sens Actuators, A 2001;78:221–227.

[36] Young M. Biotoxin detection using parallel separation arrays [Online]. Available: http://www.sandia.gov/microfluidics/research/biotox.php [Accessed], 2015.

[37] Zeng X, Mao XI, Greif R, Russo R. Experimental investigation of ablation efficiency and plasma expansion during femtosecond and nanosecond laser ablation of silicon. Appl Phys A 2005;80:237–241.

[38] Gu D. Microchannel fabrication in PMMA based on localized heating by nanojoule high repetition rate femtosecond pulses. J Opt Soc Am 2005;13:5939–5946.

[39] Suriano R, Eatonc SM, Kiyanb R, Cerullod G, Osellamec R, Chichkovb BN, Levia M, Turria S. Femtosecond laser ablation of polymeric substrates for the fabrication of microfluidic channels. Appl Surf Sci 2011;257:6243–6350.

[40] Roberts M, Bercier P, Girault H. UV laser machined polymer substrates for the development of microdiagnostic systems. Anal Chem 1997;69:2035–2042.

[41] Waddell EA, Kramer GW. UV laser micromachining of polymers for microfluidic applications. JALA 2002;7:78–82.

[42] Baudach S, Krüger J, Kautek W. Ultrashort pulse laser ablation of polycarbonate and polymethylmethacrylate. Appl Surf Sci 2000;154-155:555–560.

[43] De Marco C, Suriano R, Turri S, Levi M, Ramponi R, Cerullo G, Osellame R. Surface properties of femtosecond laser ablated PMMA. ACS Appl Mater Interfaces 2010;2:2377–2384.

[44] Deepak Kallepalli LN, Soma VR. Fabrication and optical characterization of microstructures in poly(methylmethacrylate) and poly(dimethylsiloxane) using femtosecond pulses for photonic and microfluidic applications. Appl Opt 2010;49:2475–2489.

[45] Krüger J, Madebach H, Urech L, Lippert T, Wokaun A, Kautek W. Femto- and nanosecond laser treatment of doped polymethylmethacrylate. Appl Surf Sci 2005;247:406–411.

[46] Liu ZQ, Yi XS. Coupling effects of the number of pulses, pulse repetition rate and fluence during laser PMMA ablation. Appl Surf Sci 2000;165:303–308.

[47] Pfleging W, Hanemann T, Torge M, Bernauer W. Rapid fabrication and replication of metal, ceramic and plastic mould inserts for application in microsystem technologies. ProQuest Sci J 2003;217:53–63.

[48] Pugmire DL, Haasch R, Tarlov MJ, Locascio LE. Surface characterization of laser-ablated polymers used for microfluidics. Anal Chem 2002;74:871–878.

[49] Sowa S, Tamaki T. Symmetric waveguides in poly(methylmethacrylate) fabricated by femtosecond laser pulses. J Opt Soc Am 2006;14:291–297.

[50] Suriyage N, Iovenitti P, Harvey EC. Fabrication, measurement and modeling of electroosmotic flow in micromachined polymer microchannels. BioMEMS Nanotechnol 2004;5275:149–160.

[51] Wolynski A, Hermann T, Mucha P, Haloui H, L'huillier J. Laser ablation of CFRP using picosecond laser pulses at different wavelengths from UV to IR. Proceedings of the Sixth International WLT Conference on Lasers in Manufacturing; Munich, Germany; 23–26 May 2011.

[52] Teixidor D, Thepsonthi T, Ciurana J, Özel T. Nanosecond pulsed laser micromachining of PMMA-based microfluidic channels. J Manuf Processes 2012;14(4):435–442.

[53] Stieglitz T, Beutel H, Schuettler M. Micromachined, polyimide-based devices for flexible neural interfaces. Biomed Microdevices 2000;2:283–294.

[54] Chu CC. Polyesters and polyamides. In: *Concise Encyclopedia of Medical & Dental Materials*. New York: Pergamon Press; 1990. p 261–271.

[55] He F, Cheng Y. Femtosecond laser micromachining: frontier in laser precision micromachining. Chin J Lasers 2007;05:TN249.

[56] Huang H, Guo Z. Ultra-short pulsed laser PDMS thin-layer separation and micro-fabrication. J Micromech Microeng 2009;19:055007.

[57] Wild MJ, Gillner A, Poprawe R. Locally selective bonding of silicon and glass with laser. Sens Actuators, A 2001;93:63–69.

[58] Bauer HF, Chiha M. Axisymmetric oscillation of a viscous liquid covered by an elastic structure. J Sound Vib 2006;281:835–847.

[59] Lugscheider E, Reisgen U, Remmel J, Sigismund E, Martinelli AE, de Almeida Buschinelli AJ, do Nascimento RM. *Brazing Metals to Ceramics Mechanically Metalized with Titanium*. München: Werkstoffwoche Materialica; 2002.

[60] Haferkamp H, Burmester I, Frohmann A, Kreutzburg K. Soldering of ceramic and metal with laser irradiation. In: *High Temperature Brazing and Diffusion Welding*. DVS-Berichte; 1998.

[61] Shanmugarajan B, Padmanabham G. Fusion welding studies using laser on Ti-SS dissimilar combination. Opt Lasers Eng 2012;50:1621–1627.

[62] Mathieu A, Shabadi R, Deschamps A, Suery M, Mattei S, Grevey D, Cicala E. Dissimilar material joining using laser, Al to SS using zinc-based filler wire. Opt Laser Technol 2007;39:652–661.

[63] Teixidor D, Orozco F, Thepsonthi T, Ciurana J, Rodríguez CA, Özel T. Effect of process parameters in nanosecond pulsed laser micromachining of PMMA-based microchannels at near-infrared and ultraviolet wavelengths. Int J Adv Manuf Technol 2013;67:1651–1664.

[64] Criales LE, Orozco PF, Medrano A, Rodríguez CA, Özel T. Effect of fluence and pulse overlapping on fabrication of microchannels in PMMA/PDMS via UV laser micromachining: modeling and experimentation. Mater Manuf Processes 2015;30(7):890–901.

[65] Yuan D, Das S. Experimental and theoretical analysis of direct-write laser micromachining of polymethyl methacrylate by CO_2 laser ablation. J Appl Phys 2007;101:024901–024906.

[66] Thepsonthi T, Milesi N, Özel T. Design and prototyping of micro-needle arrays for drug delivery using customized tool-based micro-milling process, Proceedings of the First International Conference on Design and Processes for Medical Devices, May 2–4, 2012, Brescia, Italy; 2012.

[67] Kleine KF, Whitney B, Watkins KG. Use of fiber lasers for micro cutting applications in medical device industry. 21st International Congress on Applications of Lasers and Electro-Optics; 2002.

[68] Muhammad N, Whitehead D, Boor A, Li L. Precision machine design. Comparison of dry and wet fibre laser profile cutting of thin 316L stainless steel tubes for medical device applications. J Mater Process Technol 2010;210:2261–2267.

[69] Baumeister M, Dickman K, Hoult T. Fiber laser micro-cutting of stainless steel sheets. J Appl Phys A 2006;85:121–124.

[70] Scintilla LD, Sorgente D, Tricarico L. Experimental investigation on fiber laser cutting of Ti6Al4V thin sheet. J Adv Mater Res 2011;264–265:1281–1286.

[71] Powell J, Al-Mashikhi SO, Voisey KT. Fibre laser cutting of thin section mild steel: an explanation of the 'striation free' effect. Opt Lasers Eng 2011;49:1069–1075.

[72] Teixidor D, Ciurana J, Rodriguez CA. Dross formation and process parameters analysis of fibre laser cutting of stainless steel thin sheets. Int J Adv Manuf Technol 2014;71:9–12, 1611–1621.

[73] Yilbas BS, Abdul Aleem BJ. Dross formation during laser cutting process. J Phys D: Appl Phys 2006;39:1451–1461.

[74] Tani G, Tomesani L, Campana G, Fortunato A. Quality factors assessed by analytical modelling in laser cutting. Thin Solid Films 2004;453, 454:486–491.

[75] Schuöcker D, Aichinger J, Majer R. Dynamic phenomena in laser cutting and process performance. Physics Procedia 2012;39:179–185.

[76] Kathuria YP. Laser microprocessing of metallic stent for medical therapy. J Mater Process Technol 2005;170:545–550.

[77] Pfeifer R, Herzog D, Hustedt M, Barcikowski S. Pulsed Nd:YAG laser cutting of NiTi shape memory alloys—Influence of process parameters. J Mater Process Technol 2010;210:1918–1925.

[78] Shanjin L, Yang W. An investigation of pulsed laser cutting of titanium alloy sheet. Opt Lasers Eng 2006;44:1067–1077.

[79] Yung KC, Zhu HH, Yue TM. Theoretical and experimental study on the kerf profile of the laser micro-cutting NiTi shape memory alloy using 355 nm Nd:YAG. Smart Mater Struct 2005;14:337–342.

[80] Muhammad N, Whitehead D, Boor A, Oppenlander W, Liu Z, Li L. Picosecond laser micromachining of nitinol and platinum–iridium alloy for coronary stent applications. Appl Phys A 2012;106:607–617.

[81] Huang H, Zheng HY, Lim GC. Femtosecond laser machining characteristics of nitinol. Appl Surf Sci 2004;228:201–206.

[82] Scintilla LD, Tricarico L. Experimental investigation on fiber and CO_2 inert gas fusion cutting of AZ31 magnesium alloy sheets. Opt Laser Technol 2013;46:42–52.

5

MACHINING APPLICATIONS

Tuğrul Özel

Department of Industrial and Systems Engineering, School of Engineering, Rutgers University, Piscataway, NJ, USA

Elisabetta Ceretti

Department of Mechanical Engineering and Industrial Engineering, University of Brescia, Brescia, Lombardy, Italy

Thanongsak Thepsonthi

Department of Industrial Engineering, Burapha University, Chon Buri, Thailand

Aldo Attanasio

Department of Mechanical Engineering and Industrial Engineering, University of Brescia, Brescia, Lombardy, Italy

5.1 INTRODUCTION

Metals have been the primary materials in the past for machining applications due to their superior mechanical properties. Metallic biomedical implants work under complex mechanical load in a salty environment, and it demands a high corrosion resistance of the material. Corrosion of the implants may release particles or metallic ions in the body and cause premature implant failure; besides, adverse reactions may occur. Biocompatible metal materials (biometals) such as stainless steel, cobalt-chromium-molybdenum alloys, titanium, and titanium-nickel alloys have been used [1, 2] in orthopedic implants, fixation systems, and many other medical implants with several applications (see Table 5.1 and Figure 5.1).

Biomedical Devices: Design, Prototyping, and Manufacturing, First Edition.
Edited by Tuğrul Özel, Paolo Jorge Bártolo, Elisabetta Ceretti, Joaquim De Ciurana Gay, Ciro Angel Rodriguez, and Jorge Vicente Lopes Da Silva.
© 2017 John Wiley & Sons, Inc. Published 2017 by John Wiley & Sons, Inc.

TABLE 5.1   Medical Implants and Alloy Materials Used

Medical Field	Implant Type	Alloy Material
Cardiovascular	Stents and valves	SS-316L; Co-Cr-Mo; Ti, Ti-6Al-4V
Craniofacial	Plates and screws	316L SS; CoCrMo; Ti; Ti6Al4V
Dental	Orthodontic wires	SS-316L; Co-Cr-Mo; Ni-Ti; Ti-Mo
Orthopedic	Joint replacements	SS-316L; Ti; Ti-6Al-4V; Ti-6Al-7Nb; Co-Cr-Mo
Spinal	Fixation plates, screws, pins	SS-316L; Ti; Ti-6Al-4V

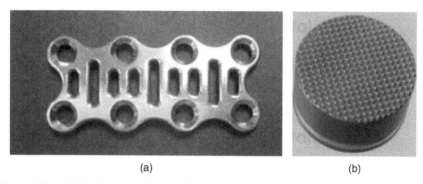

(a)   (b)

Figure 5.1   (a) Titanium alloy Ti-6Al-4V spinal fixation plate and (b) SS316L stainless steel bone crusher as produced with micromilling.

Biometals used in orthopedic implants mainly include surgical-grade stainless steel, cobalt-chromium alloys, titanium and nickel-titanium alloys, among others.

In general, stainless steel is not suitable for a permanent implant because of its poor fatigue strength, its liability to undergo plastic deformation, and poor corrosion resistance. For this reason, stainless steel is used only for nonpermanent implants such as internal fixation devices for fractures. Most commonly, permanent implants are made of cobalt-based alloys (Co-Cr-Mo and Co-Cr-Ni) or Ti alloys. These alloys are more corrosion resistant; particularly, Co alloys generate a durable chromium-oxide surface layer (so-called passivation layer). Despite good corrosion resistance, chromium and nickel are known as carcinogens; in addition, their elastic modulus (221 GPa) is roughly 10 times the stiffness of cortical bone. It is necessary to consider that the regenerative and remodeling processes in bones are directly triggered by loading; that is, bone subjected to loading or stress regenerates and bone not subjected to loading results in atrophy. Thus, the effect of a much stiffer bone implant is to reduce the loading on bone resulting in the phenomenon called "stress shielding." This phenomenon can lead to resorption of the bone and to the loss of the implant. Those findings have led to the use of titanium alloy for prosthetic devices (specifically for cemented ones).

Titanium alloys are preferred typically for medical implants because of light weight, high strength, and biocompatibility. Also, titanium implants are compatible

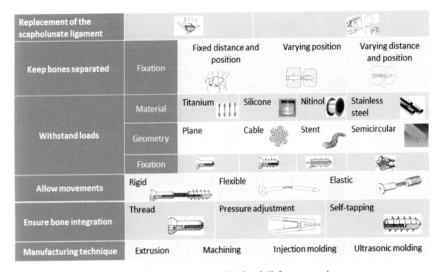

Replacement of the scapholunate ligament					
Keep bones separated	Fixation	Fixed distance and position	Varying position		Varying distance and position
Withstand loads	Material	Titanium	Silicone	Nitinol	Stainless steel
	Geometry	Plane	Cable	Stent	Semicircular
	Fixation				
Allow movements	Rigid		Flexible		Elastic
Ensure bone integration	Thread		Pressure adjustment		Self-tapping
Manufacturing technique	Extrusion		Machining	Injection molding	Ultrasonic molding

Figure 2.8   See text page 34 for full figure caption.

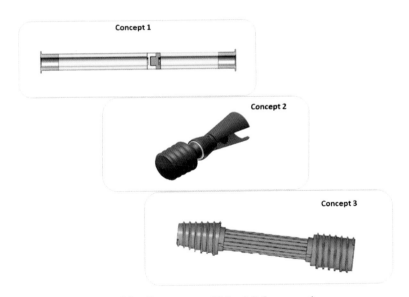

Concept 1

Concept 2

Concept 3

Figure 2.9   See text page 34 for full figure caption.

Biomedical Devices: Design, Prototyping, and Manufacturing, First Edition.
Edited by Tuğrul Özel, Paolo Jorge Bártolo, Elisabetta Ceretti, Joaquim De Ciurana Gay,
Ciro Angel Rodriguez, and Jorge Vicente Lopes Da Silva.
© 2017 John Wiley & Sons, Inc. Published 2017 by John Wiley & Sons, Inc.

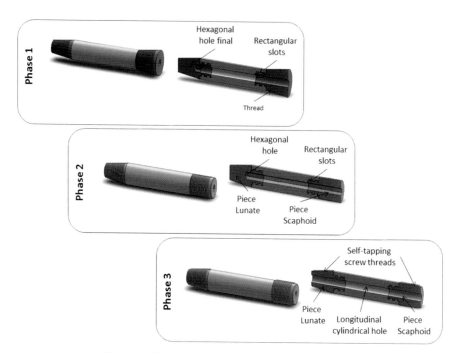

Figure 2.10 See text page 35 for full figure caption.

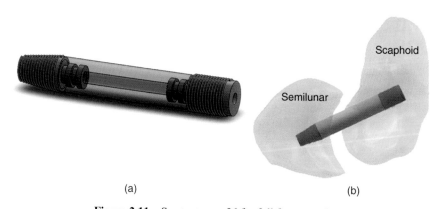

(a)   (b)

Figure 2.11 See text page 36 for full figure caption.

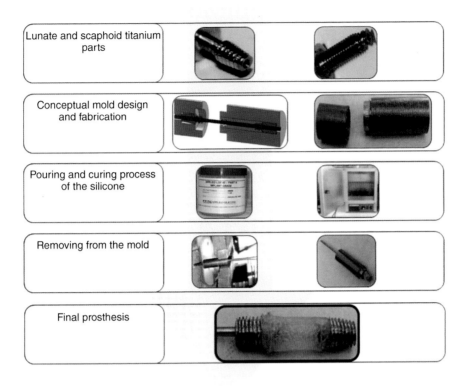

Lunate and scaphoid titanium parts		
Conceptual mold design and fabrication		
Pouring and curing process of the silicone		
Removing from the mold		
Final prosthesis		

Figure 2.12   See text page 37 for full figure caption.

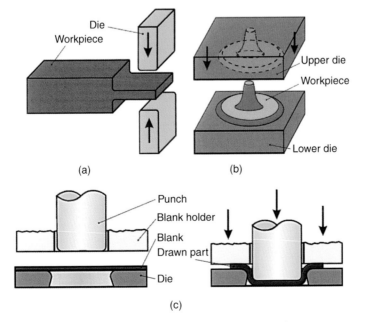

(a)

Die
Workpiece

Upper die
Workpiece
Lower die

(b)

Punch
Blank holder
Blank
Drawn part
Die

(c)

Figure 3.1   See text page 50 for full figure caption.

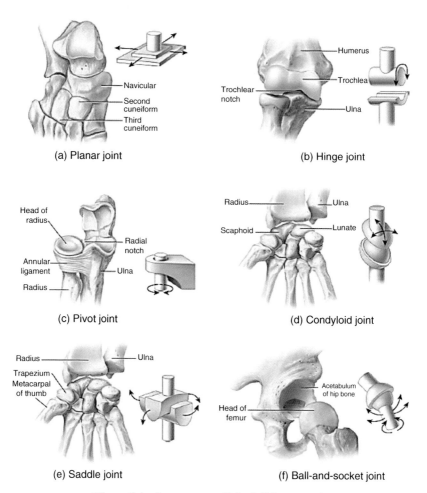

(a) Planar joint

(b) Hinge joint

(c) Pivot joint

(d) Condyloid joint

(e) Saddle joint

(f) Ball-and-socket joint

Figure 3.4   See text page 60 for full figure caption.

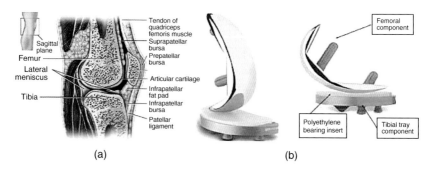

(a)

(b)

Figure 3.6   See text page 61 for full figure caption.

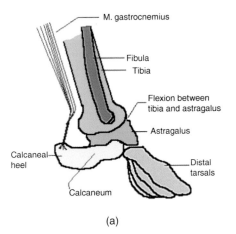

(a)

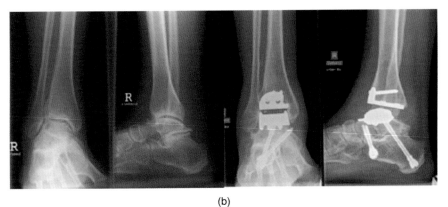

(b)

Figure 3.7 See text page 62 for full figure caption.

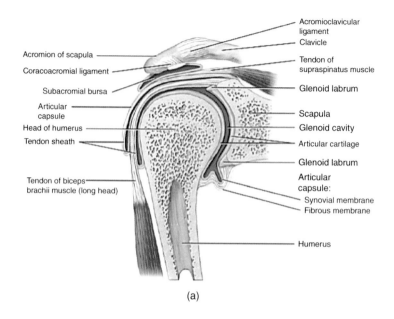

Acromioclavicular ligament
Clavicle
Tendon of supraspinatus muscle
Glenoid labrum
Scapula
Glenoid cavity
Articular cartilage
Glenoid labrum
Articular capsule:
Synovial membrane
Fibrous membrane
Humerus

Acromion of scapula
Coracoacromial ligament
Subacromial bursa
Articular capsule
Head of humerus
Tendon sheath
Tendon of biceps brachii muscle (long head)

(a)

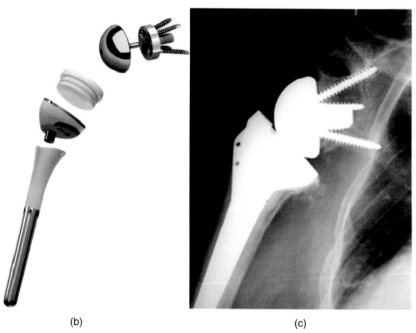

(b)

(c)

Figure 3.10 See text page 65 for full figure caption.

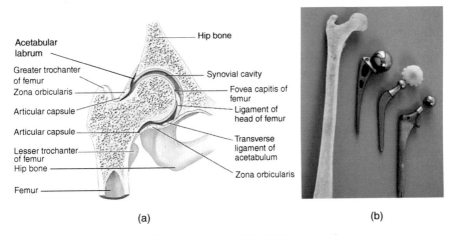

(a)

(b)

Figure 3.11 See text page 66 for full figure caption.

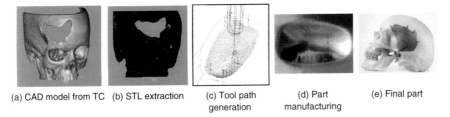

(a) CAD model from TC   (b) STL extraction   (c) Tool path generation   (d) Part manufacturing   (e) Final part

Figure 3.16 See text page 70 for full figure caption.

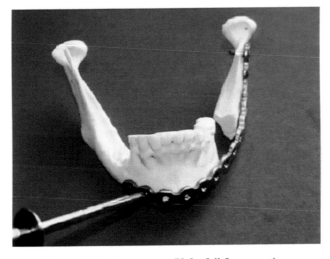

Figure 3.17 See text page 72 for full figure caption.

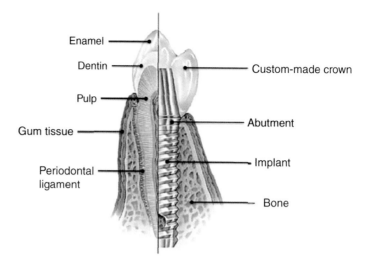

Figure 3.18 See text page 72 for full figure caption.

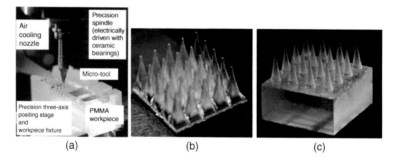

Figure 5.3 See text page 110 for full figure caption.

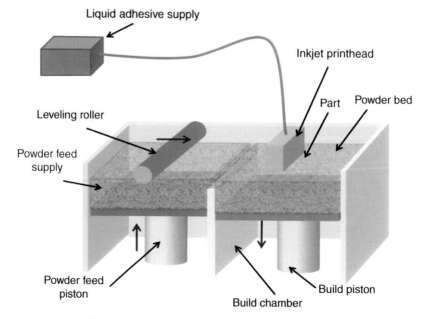

Figure 6.1 See text page 126 for full figure caption.

with magnetic resonance imaging and computed tomography imaging procedures; therefore, they do not interfere with the procedures if the patient needs them after the implant is made. Titanium and its alloys have also been demonstrated to be more corrosion resistant than cobalt-chrome alloys because of the formation of titanium oxide, which is in fact a ceramic, on the surface. Moreover, these metals have been considered to be the most biocompatible among metal materials. However, titanium does bond directly to bone, which can result in the loosening of the implant. One approach to improving implant lifetime is to coat the metal surface with a bioactive material that can promote the formation and adhesion of hydroxyapatite, the inorganic component of natural bone. Bioglass coatings on Ti implants further improve the biocompatibility of these implants [3].

The glasses are based on mixtures of the oxides of silicon, sodium, potassium, calcium, and magnesium. Also, the glasses become soft at the processing temperature, which is well below the melting point of the Ti alloy. Thus, they flow to uniformly coat the Ti surface. These coatings develop a layer of hydroxyapatite on their outer surface upon exposure to simulated body fluid. Another important issue connected with the use of metal materials in implant generation is the dangerous particles that result from wear debris. Also, in this case, it has been verified that a bioglass coating can reduce the problem.

Some of the specific mechanical properties of the alloys used in joint replacements and their microstructure are summarized in Table 5.2 [2, 4].

Pure titanium and Ti-6Al-4V are the main materials used in the dental field and in the surgical field. Ti-6Al-7Nb, which has been developed for surgical implants, is also attractive for dental applications.

TABLE 5.2   Mechanical Properties of Alloys Used in Joint Replacements

Alloy Material	Microstructure	Tensile Strength (MPa)	Modulus (GPa)
Ti (pure titanium)	{α}	785	105
Ti-Zr	Cast {α'/β}	900	1
Co-Cr alloys		655–1896	0.9–253
Co-Cr-Mo	{Austenite (fcc) + hcp}	600–1795	8.3–230
Ti-6Al-4V	{α'/β}	960–970	0.4
Ti-6Al-7Nb	{α/β}	1024	105
Ti-5Al-2.5Fe	{α/β}	1033	110
Ti-13Nb-13Zr	{α'/β}	1030	79
Ti-15Mo-5Zr-3Al	{Metastable β}	882–975	75
Ti-12Mo-6Zr-2Fe	{Metastable β}	1060–1100	74–85
Ti-15Mo-5Zr-3Al	{Metastable β}	882–975	75
SS-316L	{Austenite}	465–950	200
Ti-15Mo-2.8Nb-3Al	{Metastable β}	812	82
Ti-35Nb-5Ta-7Zr	{Metastable β}	590	55
Ti-35Nb-5Ta-7Zr-0.4O (TNZTO)	{Metastable β}	1010	66

TABLE 5.3   Biocompatibility of Biomaterials Including Metal Alloys

Biomaterial	Patterns	Biocompatibility
Stainless steel, Co–Cr alloy, titanium, titanium alloys	Intervend osteogenesis Contact osteogenesis	Biotolerant materials Bioinert materials
Bonding osteogenesis	Bioglass, Ceravital tricalcium phosphate, hydroxyapatite	Bioactive materials

Titanium alloys show the greatest biocompatibility among metallic materials for biomedical applications (see Table 5.3). However, they are grouped into bioinert materials as well as ceramics such as alumina and zirconia, judging from the pattern of osteogenesis as stated earlier, and its biocompatibility is inferior to that of phosphate calcium or hydroxyapatite, which is grouped into bioactive materials. Therefore, bioactive surface treatment (bioactive surface modification) is in general applied to titanium alloys for biomedical applications in order to improve their biocompatibility [3, 5].

5.2   MACHINABILITY OF BIOCOMPATIBLE METAL ALLOYS

Orthopedic devices are designed to conform to the complex shape of bones and joints; therefore, the machining of these parts is also complex. Devices machined from bar stock require a lot of material to be removed, resulting in an expensive process because of the low machinability rating of many of the materials involved. As a result, some parts are cast to near net shape, which often requires fixturing that is complex and expensive. Another issue that adds to the complexity of machining is the tight tolerances required—50 μm or less for most devices.

Machinability of biocompatible metal alloys, such as stainless steel, titanium alloys, and titanium-nickel alloys, is considered highly difficult due to their low thermal conductivity and diffusivity, high rigidity and low elasticity modulus, and high chemical reactivity at elevated temperatures; and resultant machining-induced surface quality and structural surface integrity may not be acceptable at the readiness levels suitable for biomedical applications as reviewed in the literature by several peer research groups [6, 7].

As a group of shape memory alloy with ample biomedical applications, the nickel-titanium alloys based on nitinol (Ni-Ti) is also widely used because of their shape memory-based actuating property that can be employed in catheters, forceps, biopsy, and surgical devices. Machining these alloys was found to be very difficult, owing to their high ductility and work-hardening behavior [8]. Weinert and Petzoldt suggested that for these materials, the tool wear is high whether the feed rate and cutting speed are high or low; therefore, increasing the material removal rate while taking care of the surface quality by rapid tool changing was recommended. Micromilling was proposed as a favorable method to machine these materials, and using a minimum quantity of lubrication was suggested for increased surface quality [9]. Micromilling at a cutting speed of 33 m/min, feeds of 6–30 µm/tooth, depths of cut of 10–100 µm, and a width of cut of 250 µm, a high feed rate, and a relatively high width of cut was found to form better chips, which extended tool life and enhanced the workpiece quality. In addition to minimum quantity lubrication, tool coating was suggested rather than using uncoated tools, but multilayer TiCN/TiAlN- or TiCN/TiN-coated cemented carbide tools were found to provide better results compared with PCD or CBN tools in turning and drilling processes in terms of workpiece quality and tool costs [7].

Once machining induced properties are controlled, these types of biocompatible metallic materials can be extremely useful for the biomedical industry, because of their improved surface quality and surface integrity properties. They can be used in biomedical devices such as stents, dental implants, orthopedic implants, and other devices, and their high biocompatibility with the human body is a significant concern.

The surface integrity of the final biomedical part is highly crucial in machining processes. In most applications, having the smoothest possible surface is desired, especially when strength under static loading and fatigue life of a medical device component is important. However, in some cases, having a rougher surface can be preferred typically in some medical device applications or in the biomedical field generally.

Machining processes induce and affect various surface integrity attributes on the finished parts [7]. These can be grouped as (1) topography characteristics such as textures, waviness, and surface roughness, (2) mechanical properties affected such as residual stresses and hardness, and (3) metallurgical state such as microstructure, phase transformation, grain size and shape, inclusions, etc. These alterations of surface are considered in five groups: mechanical, thermal, metallurgical, chemical, and electrical properties.

Machining can cause many defects on the surfaces produced since it is a material removal process. Main forms of defects are surface drag, material pull-out/cracking, feed marks, adhered material particles, tearing surface, chip layer formation, debris of microchips, surface plucking, deformed grains, surface cavities, slip zones, laps (material folded onto the surface), and lay patterns [7].

The cutting parameters can affect these defects to some degree, which need to be controlled. Feed marks are effective in machining, but their severity can be altered by varying and optimizing the feed rate. Cutting speed values can affect the amount of microchip debris on the surface, and material plucking, tearing, dragging, and smearing can be affected by depth of cut among other parameters. During machining of nickel and titanium alloys, such problems can be problematic, so optimizing the cutting conditions is essential [7].

There are many surface defects that can be found in machining processes, especially when investigated in high precision. The main surface defects are considered to be feed marks, chip redeposition to the surface, and grain deformations, since these are the ones in the biggest scale among the surface defects. Also, plucking of particles from the surface and their redeposition to the surface create two different defects, whereas these particles can also cause dragging and tearing defects on the next pass from the surface. Adjusting cutting parameters according to these defects is very hard, and even then, a complete elimination is not possible [7].

Many workpiece materials include carbide particles in their structure. Also, many coating materials involve some carbide in them. As the tool wears, and the workpiece is machined, these carbide particles are sometimes removed from the machined surface or the tool and get stuck on the workpiece surface. This phenomenon is called carbide cracking, and it causes a sudden increase in the shear stress during cutting that leads to surface cavities due to plucking. This process causes residual cavities and cracks to be formed inside the machined surface, causing even further problems [7].

Both titanium and nickel alloys are prone to carbide cracking where the existence of crack locations decreases the fatigue life of the material substantially. These materials are strengthened by carbides such as titanium carbide and niobium carbide. When feed rates and depth of cut values lower than 50 µm are used to observe the possibility of good surface roughness, the carbide particles from these strengthening pieces were observed to crack from the surface and be smeared to another part of the workpiece material to create surface integrity problems. The sizes of these carbide particles were also found to be around 20 µm, comparable to the feed rate and depth of cut values. Carbide particles are unable to deform, which means they cannot be removed completely at once, leaving cavities behind, as well as causing high oscillations in forces. Hence, taking these possibilities into account, planning the machining processes accordingly for biomedical applications is extremely important [7].

5.3 SURFACES ENGINEERING OF METAL IMPLANTS

In particular, in the case of cobalt-chromium-molybdenum alloys, the metal-on-metal (MoM) implants show a higher wear resistance than the metal-on-polymer (MoP) ones, even if they cannot avoid the formation of wear debris. Furthermore, debris deriving from MoP implants usually generates inflammatory process with destruction of bone tissue (osteolysis), while MoM is rarely associated with osteolysis. This allows using larger diameter of the femoral heads in order to increase the stability [10].

The debris formation is mainly due to fatigue phenomenon and to high contact stresses on the surface. Under these cycles, the chromium and molybdenum carbides can be fractured, adding wear on the contact surfaces by indentation and abrasion of a third part. Thus, the generation of residual stresses of compression on the surfaces during machining could minimize the debris formation due to surface fatigue. The problem of Co-Cr-Mo alloys is that they are difficult to machine materials, since they have high work-hardening rate, low thermal conductivity, and the presence of hard abrasive carbides [11].

Usually, when the implants are under working conditions, the particular components are subjected to surface and subsurface alterations. These tribochemical effects lead to the formation of a nanocrystalline surface region that, during fatigue cycles, cracks, and this is the main cause of debris generation. In order to improve the surface characteristics of the particular components of the implants, several studies were performed by making heat treatment and surface coating. However, the heat treatment changes the chemical structure of the whole component, while with surface coating problems of spalling and adhesion with the new layers were presented.

Various surface modification technologies relating to titanium and titanium alloys including mechanical treatment, thermal spraying, sol-gel, chemical and electrochemical treatment, and ion implantation are applied from the perspective of biomedical applications. Wear resistance, corrosion resistance, and biological properties of titanium and titanium alloys can be improved selectively using the appropriate surface treatment techniques while the desirable bulk attributes of the materials are retained [11].

Magnesium and magnesium-based alloys are lightweight metallic materials that are extremely biocompatible and have mechanical properties that are similar to natural bone. These materials have the potential to function as an osteoconductive and biodegradable substitute in load bearing applications in the field of hard tissue engineering. However, the effects of corrosion and degradation in the physiological environment of the body has prevented their widespread application to date [12].

There has been research on the methods of improving the corrosion resistance of magnesium and its alloys for potential application in the orthopedic field. To be an effective implant, the surface and subsurface properties of the material need to be carefully selected so that the degradation kinetics of the implant can be efficiently controlled. Several surface modification techniques are presented and their effectiveness in reducing the corrosion rate and methods of controlling the degradation period are discussed. Ideally, balancing the gradual loss of material and mechanical strength during degradation, with the increasing strength and stability of the newly forming bone tissue, is the ultimate goal. If these methods prove to be successful, orthopedic implants manufactured from magnesium-based alloys have the potential to deliver successful clinical outcomes without the need for revision surgery [13].

In the case of orthopedic implants made of magnesium-calcium alloys, the main problem is that the material is quickly biodegradable in the body environment. Thus, one challenge is to control the degradation rate. It was shown that the optimization of the residual stress distribution could decrease the degradation rate, and by adjusting the machining-induced surface residual stress profile, this rate can be reduced significantly [6, 7].

5.4 WEAR AND FAILURE OF METAL IMPLANTS

The joint replacement procedure involves surgical operation for removing the diseased area and replacing the joint with a prosthesis made of biocompatible metal material. Once the prostheses is implanted in the patient, several factors may affect the performance of the metal implant, such as the interaction with the human body or the mechanical loads, of static and dynamic nature, generated during the daily movements. These could lead to the failure of the metal implant and consequently to risks for patient health. The failure modes resulting from machining processes could include high cycle fatigue cracking due to surface microcracks, wear of the surfaces in contact with relative debris formation due to undesirable subsurface quality, and bacterial annexation due to the absorption of biomolecules by the prostheses metal material that usually causes inflammation of soft and hard tissues. These modes are deeply related to the metal material utilized for the prostheses production via machining processes. For this reason, failure is one of the most important aspects of implant material behavior and directly influences the choice of metal materials and machining methods [7, 11].

In the case of temporary metal implants, such as plates and screws utilized for fixing fractured bones, the metal implant alloy needs to be bioabsorbable and biodegradable in the body environment. However, in order to permit the growth of bone tissue and make the tissue growth stable, the metal implant material should not be highly degradable during the bone development period, where the stability of the temporary implant and of the bone tissue is reported as a function of the time after the implantation. Magnesium-calcium alloys are usually used for this purpose. The failure related to this type of alloy is the rapid biodegradation when they are placed in the human body. This could result in incomplete bone formation before the complete absorption of the implants, with a catastrophic breakage due to the poor stability of the new bone tissue. This problem could be solved by controlling the biodegradation rate and avoiding the absorption of the implant, by realizing the final parts with specific surface characteristics [12].

When metallic permanent joint prostheses are considered, the materials constituting all the parts are not biodegradable. In this case, the main causes of failure are high cycle fatigue, corrosion under stress phenomena, wear and debris formation. The failure under high cycle fatigue is usually related to a wrong prostheses design. In such case, under torsional and bending momentum, a high stress concentration area could occur, leading to a premature cracking of the implant. The high level contact stresses generated between the implant parts, during the running, should cause the beginning of corrosion phenomenon [12].

When the wear particles are produced, they generate micro-cutting and micro-ploughing phenomena due to abrasions that lead to the formation of scratches and grooves on the surfaces. Moreover, agglomeration, compaction, and cold sintering of debris, due to high contact stresses and sliding conditions, generate adhesive wear.

Since the failure of metallic implants are mainly related to wear and debris formation, which are heavily affected by the surface integrity, it results in great interest to find a correlation between the surface characteristics and growing wear [12].

5.5 MICROMILLING-BASED FABRICATION OF METALLIC MICROCHANNELS FOR MEDICAL DEVICES

Special medical devices for microfluidic applications such as lab-on-chip (LOC), micropump, microvalves, and microthermal devices are diffused in the medical field.

In particular, the LOC is a device integrating functions of several laboratories (sample pretreatment, cleaning, and separation) on a single chip. An LOC is composed by channels with different sizes and geometries, accurately designed for handling a very small fluid volume (less than picoliters). DNA analysis, enzymatic analysis, and clinical pathology are some of the application areas of LOC. The main advantages introduced by LOC are the cost reduction, the low consumption of fluid and the faster analysis and response times.

Different manufacturing processes can be used for realizing the microchannels on the chip, amongst them photolithography, etching, additive manufacturing, laser, electroplating, embossing and, recently, machining, that is, micromilling. Although these micro-manufacturing processes allow obtaining features characterized by high geometrical accuracy, precision, and low surface roughness, often they are not able to guarantee the needed precision when microscale is considered. In terms of machining-related issues, quality of channels geometry, precision of channel dimensions, burrs at the edges of the channels, and surface roughness at the bottom and side walls are some of the aspects to be considered for having good quality analyses [14].

The Micromilling process is one of the mechanical micromachining processes. It is a scale-down version of the conventional end milling process. Micro-end-milling is typically used to produce miniature parts that are not axially symmetric and have many features, such as holes, slots, pockets, and even three-dimensional surface contours. Micromilling is still a tool-based material removal process; hence, it very much relies on the performance of the micromilling tools. Also, this happens to be a barrier limiting the capability of mechanical micromachining. This problem is encountered when scaling down the tool; there are some features of the tool that cannot be reduced further. This problem affects one of the most critical tool geometry, the edge radius. A limitation of grinding technology, grain size, and edge strength limits further reduction of the edge radius. Inevitably, the microtools are normally fabricated with an edge radius of 1–5 μm [14–18]. When considering micromilling as the manufacturing process, these aspects are strongly affected by the process parameters (depth of cut, lubrication type, feed per tooth) [9, 14, 15], tool material and coating [15], and material microstructure [16]. As a consequence, it is very important to know how to select and optimize these process parameters for achieving the product quality requirements [17, 18].

The channel geometries and dimensions fabricated with micromilling are strongly influenced by the depth of cut. This is related to the high forces generated during the process that deflect the tool causing evident geometrical errors or deflected geometries [14, 15, 18]. The coolant factor has the highest influence on surface finish. Wet cutting considerably reduces the roughness in the floor of the channel. Moreover, an improvement in the burr reduction can be achieved when using coolant [14]. The roughness in the floor of the channel is also affected by the feed per tooth values. As in conventional slot milling operation, high feed values generate worst surface finishing. It is well known that also the material microstructure affects the final part quality, tool wear, and cutting force [16].

Ti-6Al-4V alloy can show different microstructures depending on recrystallization annealing time and temperature: bimodal (duplex) microstructure, containing
equiaxed primary α in a lamellar α + β matrix, which has an excellent combination
of mechanical properties; fully lamellar microstructure, having high toughness but
low ductility; fully equiaxed microstructure, with fairly good strength and ductility;
mill annealed microstructure, which is the result of cooling after plastic deformation,
without any recrystallization annealing [16, 19].

There are several responses that are used to evaluate the process performance of
micro-end-milling. Among them, burr formation, surface roughness, tool wear, cutting force, and cutting temperature are the key responses determining the success of
the process. Especially, burr formation and surface roughness are directly related to
the satisfactory level of quality produced [9, 15–18].

Burr is an undesirable projection part of a workpiece, which is produced
through manufacturing processes on an edge or a surface that lies outside the
desired geometry. Even though burr is not desirable, it is unavoidable. The only
solution is to reduce it to an acceptable degree in which a machined part can
function properly. Burr in end-milling can be categorized by the position where
it occurs. Burr formation mechanism is very complicated involving plastic and
elastic deformations, which are influenced by material properties, tool geometry, and
machining parameters. Burr in micro-end-milling of metal alloys such as titanium
alloy can be relatively large compared with the feature size as shown in Figure 5.2.
Typically, burrs are more pronounced when the down-milling cycle is more effective
than the up-milling cycle on the walls of the microchannels. It is also shown that
the increasing wear of micromilling tools affect the increase of the burrs at the
exit locations of the microchannels [15, 17, 18]. There is significant influence of

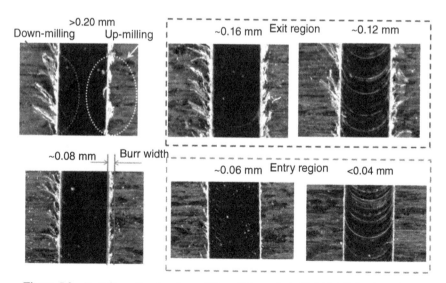

Figure 5.2 Burr formation in micromilling of channels in Ti-6Al-4V titanium alloy.

Ti-6Al-4V microstructures on burr's width and height, cutting forces, and built up edge [16]. The different behaviors are due to the different hardness and ductility of the Ti-6Al-4V microstructures. Moreover, considering only the micro-workability of this titanium alloy [19], it was demonstrated that it is strongly affected by material ductility, while hardness is a less relevant parameter [19].

The location of burr formation in medical-grade titanium alloys creates significant difficulties in terms of deburring, cleaning, and postprocessing using electropolishing, as complex features are fabricated in medical device applications (see Figure 5.2),

5.6 MACHINING-BASED FABRICATION OF POLYMERIC MICRONEEDLE DEVICES

Microneedles are innovative medical devices with the same purpose of classic hypodermic needles but fabricated on microscale often in the form of arrays in various materials. These devices aim to replace the hypodermic needles in some applications and consist of a patch with microsized needles. These microneedle array-based patches generally do not induce pain to the level of hypodermic needles since they penetrate into the skin deep enough to deliver the drug but do not reach the pain receptors. In addition, they can be applied without the help of a health-care professional, lowering the cost of delivery and improving preparedness and readiness against fighting with epidemics [20]. The basic premise of painless microneedle patches is the small individual needle size (a length of 0.5–1.5 mm and a diameter about 100–200 μm) so that each microneedle does not reach deeper than the dermis of human skin and does not agitate pain receptors but is effective in delivering the drug into the body [21]. Microneedles also offer a broad range of advantages when compared with traditional hypodermic needles [22].

Microneedle array-based patch prototypes can be developed using micromilling technology. Thepsonthi et al. [22] investigated the feasibility of directly fabricating a microneedle array-based patch prototype using micromilling of PMMA polymeric material. There are a number of design parameters for the microneedle arrays. These include a basic shape of a microneedle, the height, the base diameter (conical shape) or the base dimension (square-based pyramidal shape), horizontal and vertical spaces, and linear or circular array type. The microneedle shape is defined by considering the biomedical needs and the limitation in micromilling. The patch has to be made of biocompatible or dissolvable polymer; the needle height must be about 500 μm to 1 mm for drug storage; and the needle tips have to be sharp and penetrate easily through the skin but not fail prematurely. The micromilling process limits are related specifically to microtool sharpness and resistance of the polymeric tip during micromilling.

Due to their low thermal conductivity, polymers can exhibit thermal softening during micromilling; hence, a polymer that offers a high melting temperature and good strength is suitable for microneedle array fabrication using micromilling. In addition, an air cooling system should be employed by using an air nozzle to cool down the temperature in the polymer and also blow polymeric chips and

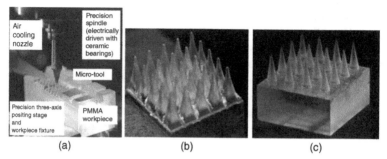

Figure 5.3   Micromilling of PMMA polymer to produce microneedle arrays. (*See color plate section for the color representation of this figure.*)

debris. Furthermore, continuous toolpaths should be designed to realize pyramidal and conical shaped microneedle arrays. Therefore, it was possible to fabricate microneedle arrays with highly sharp and burr-free tips as shown in Figure 5.3 and resultant patch prototypes as shown in Figure 5.4.

Development of microneedle arrays with various configurations such as solid, drug coated, hollow, or dissolvable type drug delivery patches using various prototyping and fabrication processes is a topic of current research studies bringing together the inter- and cross-disciplinary team of researchers from life sciences and engineering fields.

5.7   A CASE STUDY: MILLING-BASED FABRICATION OF SPINAL SPACER CAGE

The use of bioabsorbable implants in spine surgery is expanding at a rapid pace. These implants are mimicking the roles of traditional metallic devices and are demonstrating similar efficacy in terms of maintaining stability and acting as carriers for grafting substances. Biomechanical studies have demonstrated their ability to stabilize effectively a degenerative cervical and lumbar motion segment.

The objectives and assumptions of this case study are to develop the design and manufacturing of a spinal cage, starting from the material selection, product design, product finite element analysis (FEA), premanufacturing simulation and final product fabrication, to design the spinal cage which will improve the conjoint with vertebral endplate, to provide FEA for selecting better cage concerning the pattern of pyramidal teeth in superior surface. The assumptions made include the following: (i) the cage design is not specified in a certain level of cervical segment and has no concern for anterior or exterior procedures and (ii) FEA will not provide shear, flexion and extension, lateral bending, and torsion analysis; compression is the only element of concern.

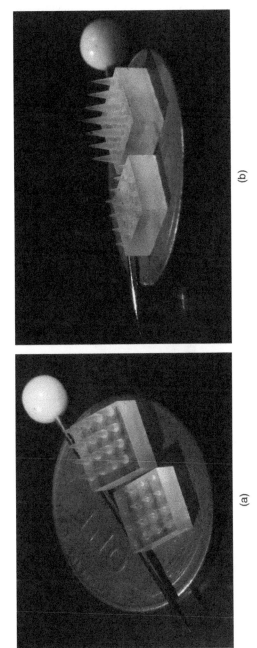

(a)

(b)

Figure 5.4 Microneedle array-based patch prototypes produced with micromilling.

5.7.1 Degenerative Disc Disease

Degenerative disc disease (DDD) is one of the most frequently encountered spinal disorders (Figure 5.5). Lumbar and cervical disc degeneration is commonly seen. Cross-sectional studies have shown that more than half of the middle-aged population demonstrated radiological or pathological evidence of cervical spondylosis. Cervical spondylosis is often asymptomatic; however, 10–15% of individuals have associated root or cord compression [23]. In the cervical spine, DDD can result in significant pain, instability, and radiculopathy and myelopathy. Several causes have been cited as the possible source of these symptoms, including the loss of disc space height, loss of foraminal volume, disc bulging, or protruding osteophytes causing neural compression. With regard to the lumbar spine, symptomatic disc degeneration is believed to be a common cause of chronic lower back pain [24].

At present, a wide array of treatment options, operative or nonoperative, for DDD is available. These treatments include anti-inflammatories, exercise, weight loss, physical therapy, discectomy with or without fusion, intradiscal electrothermal

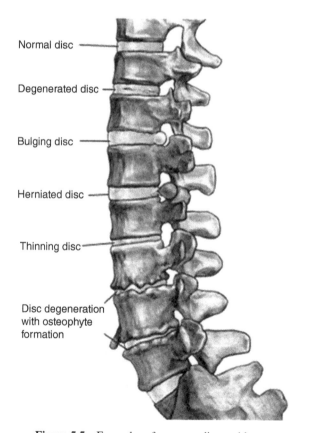

Figure 5.5 Examples of common disc problems.

therapy, prosthetic disc nucleus device, disc arthroplasty, or bioengineered nucleus pulposus replacement [24]. However, over the past 50 years, spinal fusion has generally become the standard surgical care for numerous pathologic conditions of the spine including DDD, and typically there are two types of surgical instruments for curing degenerative fusion: (1) fixation, plates, screws, rods and (2) intervertebral spacers.

Each of them has its own characteristics, but this study focuses on the intervertebral spacers through material selection, prototype design, optimization, and fabrication.

5.7.2 Intervertebral Spinal Spacers

Different design solutions were developed by considering the following requirements: the cage has to have several sagittal profiles, a variety of heights, and specific angle implants to provide treatment flexibility; the presence of superior and inferior teeth with pyramid or serrated shape to avoid implant migration; the design of a central hole for containing autograft material. Two different spinal spacer cages were designed (as shown in Figure 5.6), with different surface method and teeth pattern method (as reported in Table 5.4).

The goal of the new design presented herein was to improve the spinal cage implants built in polymeric material, available in the market due to a several complications associated with them, such as low bone growth, migration, spine

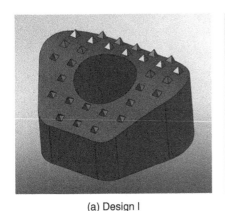 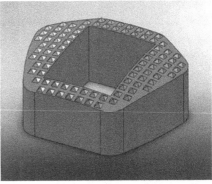

(a) Design I   (b) Design II

Figure 5.6 Spinal spacer cage designs (Rutgers Manufacturing Automation Laboratory).

TABLE 5.4 Spinal Spacer Cage Design Methods

Part	Surface Method	Teeth Pattern Method
1	Extruded cut	Linear pattern
2	Surface sweep	Curve pattern/combine

stabilization, and soft tissue formation, among others. The new cage design was generated considering some requirements: biocompatibility/biodurability, bone fusion, spinal column stabilization, stable placement, easy insertion, customizable, radiopacity, and low cost. There are basically two types of spinal cage implants: the artificial disc and fusion product. This study was only concerning the PEEK fusion spinal cages, due to good compromise between mechanical and biocompatibility characteristics of that material and the relatively higher simplicity of fusion products [25–27].

The FEA was utilized with an aim to compare two cages with different teeth densities and two different loads on each of the cages (100 N and 160 N), and to analyze the result and identify a better design. Due to the symmetry of the cage, the simulations were performed on just half models of the cage (Figure 5.7).

According to teeth density and loading, the maximum compression in the teeth was 11.9 N. Since there would be a small deformation in the top of the cage, we assume that there was no plastic deformation in this study. The bottoms of cages were set as constraints fixed in three axes during the analysis. Figure 5.8 shows the distribution of the load and constraints once the simulation was set in FEA software. The spinal cages were meshed with four-node tetrahedral mesh, as shown in Figure 5.9. Based on

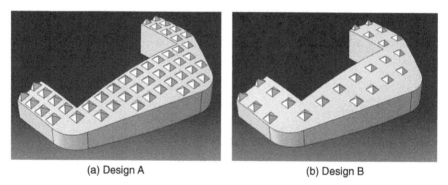

(a) Design A   (b) Design B

Figure 5.7 Two design models used in finite element analysis.

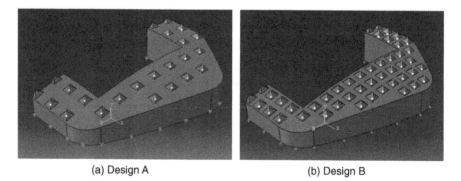

(a) Design A   (b) Design B

Figure 5.8 Boundary conditions and loads for the two designs.

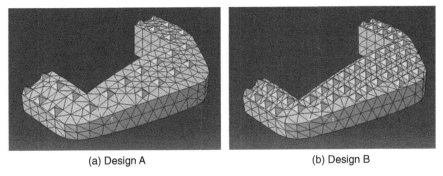

(a) Design A   (b) Design B

Figure 5.9   Finite element mesh for the two designs (Rutgers Manufacturing Automation Laboratory).

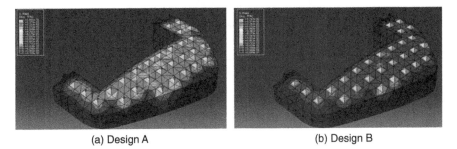

(a) Design A   (b) Design B

Figure 5.10   (a) Design A with 160 N load on 59 teeth cage and (b) Design B with 100 N load on 59 teeth cage (Rutgers Manufacturing Automation Laboratory).

the FEA, the result in Figure 5.10 shows that no deformation occurred in the design of 59 teeth cage. Therefore, we can conclude that the optimal design may be the design with the number of teeth between 42 and 59.

5.7.3   Prototype Fabrication Using Milling Process

Computer numerical control (CNC) milling process is utilized in the prototyping phase; considering the material cost, PEEK was replaced by PVC, since it has similar polymer properties. The cage size was not specified for a particular vertebrae; thus, the reasonable stock 50.75 mm × 77 mm × 50.75 mm was selected for the fabrication.

The toolpath strategies selection was critical for the prototyping purpose for several reasons. It could affect the teeth shape resulting in geometric error, rough surface, long machining time, short tool life and high cost. Figures 5.11 shows the toolpaths used in the machining simulation when a ball end mill with 0.125 in tool diameter was employed. The selection of a suitable tool not only contributes to an exquisite milling surface, but greatly enhances efficiency and reduces the cost. In addition, the proper tool selection also considers the tool loading time and machining time. In this case, a

Figure 5.11   Surface obtained from the tool path strategy (Rutgers Manufacturing Automation Laboratory).

TABLE 5.5   Tools Utilized in Milling Process

Tool Magazines		
Tool Size (in.)	Tool Type	Steps
1	End mill	Body contour lumen
0.125	Ball end mill	Rough machine for upper and bottom
0.09375	Ball end mill	Finishing machine for upper and bottom

1-in. end mill tool was utilized for the contour and lumen. Table 5.5 reports the tools adopted as a function of the different steps needed for the prototype manufacturing.

The spinal cage prototype was fabricated by a three-axis CNC machine by following the automated milling process sequence given in Figure 5.12. The first step was profile milling. This was performed on a CNC milling machine by means of a flat end mill with a diameter of 1 in. The depth of the process was 37 mm and the process parameters were a spindle speed of 2000 rpm and a feed rate of 100 mm/min. In this case, only finishing profile milling was selected, and it was due to the low hardness polymer property. The second step was hole drilling. This was also performed on a CNC milling machine by means of flat end mills with a diameter of 1 in. The depth of the process was 40 mm and the process parameters were a spindle speed of 1600 rpm and a feed rate of 800 mm/min. The hole was drilled using a peck drilling cycle to facilitate chip evacuation. The depth of the hole was 52 mm. The third step was the rough milling of the upper surface. This was performed on a CNC milling machine following the Y-parallel strategy, and by means of a two-flute ball end mill with a diameter of 0.25 in. The process parameters were a spindle speed of 2000 rpm and a feed rate of 1000 mm/min. The top part was machined with an allowance of 2 mm. Then, the finish milling of the upper surface was performed on a CNC milling machine following the toolpath strategy. The tool was a four-flute ball end mill with a diameter of 0.09375 in. The process parameters were a spindle speed of 3000 rpm and a feed rate of 300 mm/min. In order to avoid misalignment when

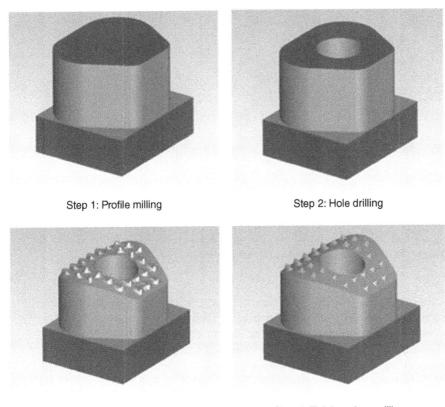

Step 1: Profile milling   Step 2: Hole drilling

Step 3: Rough surface milling   Step 4: Finish surface milling

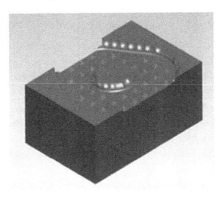

Step 5: Cage bottom fabrication simulation

Figure 5.12 Steps in automated milling process (Rutgers Manufacturing Automation Laboratory).

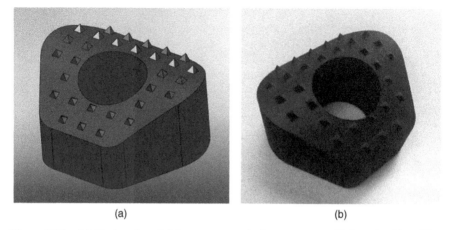

(a)   (b)

Figure 5.13   (a) Designed and (b) prototype spinal spacer cage fabricated with milling process (Rutgers Manufacturing Automation Laboratory).

reversing the part, profile milling and hole were machined deeper than the specified cage dimension. After flipping the stock to fabricate the bottom surface, the effect resulting from misalignment would be minimized. Even though the starting point is inaccurate, there would be minimal chips left along the boundary, which could be easily cleaned up. The bottom surface had to be 15.5 mm depth surface milled, then fabricate using the same tools, methods, and sequence with the superior. Fabricated prototype using this methodology is shown in Figure 5.13.

REFERENCES

[1] Black J, Hastings G, editors. *Handbook of Biomaterial Properties*. London: Chapman & Hall, Thomson Science; 1998.

[2] Hermawan H, Ramdan D, Djuansjah JRP. Metals for biomedical applications. In: Fazel R, editor. Biomedical Engineering – From Theory to Applications. Croatia: InTech; 2011.

[3] Niinomi M. Mechanical properties of biomedical titanium alloys. Mater Sci Eng 1998;A243:231–236.

[4] Long M, Rack HJ. Titanium alloys in total joint replacement—a materials science perspective. Biomaterials 1998;19:1621–1639.

[5] Niinomi M. Recent research and development in titanium alloys for biomedical applications and healthcare goods. Sci Technol Adv Mater 2003;4:445–454.

[6] Jawahir IS, Brinksmeier E, M'Saoubi R, Aspinwall DK, Outeiro JC, Meyer D, Umbrello D, Jayal AD. Surface integrity in material removal processes: recent advances. CIRP Ann 2011;60:603–626.

[7] Ulutan D, Özel T. Machining induced surface integrity in titanium and nickel alloys: a review. Int J Mach Tools Manuf 2011;51:250–280.

[8] Weinert K, Petzoldt V. Machining of NiTi based shape memory alloys. Mater Sci Eng, A 2004;378:180–184.

[9] Weinert K, Petzoldt V. Machining NiTi micro-parts by micro-milling. Mater Sci Eng, A 2008;481-482:672–675.

[10] Liua X, Chub PK, Ding C. Surface modification of titanium, titanium alloys, and related materials for biomedical applications. Mater Sci Eng 2004;R47:49–121.

[11] Deshpande A, Yang S, Puleo D, Pienkowski D, Dillon Jr O, Outeiro J, Jawahir IS. Minimized wear and debris generation through optimized machining of cobalt–chromium–molybdenum alloys for use in metal-on-metal hip implants. Proceedings of 2012 ASME International Mechanical Engineering Congress & Exposition; June 4–8, Indiana, USA: Notre Dame.

[12] Jai Poinern GE, Brundavanam S, Fawcett D. Biomedical magnesium alloys: a review of material properties, surface modifications and potential as a biodegradable orthopaedic implant. Am J Biomed Eng 2012;2(6):218–240.

[13] Denkena B, Lucas A, Thorey F, Waizy H, Angrisani N, Meyer-Lindenberg A. In: Monteiro W, editor. Biocompatible Magnesium Alloys as Degradable Implant Materials—Machining Induced Surface and Subsurface Properties and Implant Performance. *Special Issues on Magnesium Alloys*. Croatia: InTech; 2011.

[14] Monroy-Vázquez K, Attanasio A, Ceretti E, Siller HR, Hendrichs-Troeglen NJ, Giardini C. Evaluation of superficial and dimensional quality features in metallic micro-channels manufactured by micro-end-milling. Materials 2013;6:1434–1451. DOI: 10.3390/ma6041434.

[15] Thepsonthi T, Özel T. Experimental and finite element simulation based investigations on micro-milling Ti–6Al–4V titanium alloy: effects of cBN coating on tool wear. J Mater Process Technol 2013;213(4):532–542.

[16] Attanasio A, Gelfi M, Pola A, Ceretti E, Giardini C. Influence of material microstructures in micromilling of Ti6Al4V alloy. Materials 2013;6:4268–4283. DOI: 10.3390/ma6094268.

[17] Thepsonthi T, Özel T. Multi-objective process optimization for micro-end milling of Ti–6Al–4V titanium alloy. Int J Adv Manuf Technol 2012;63(9):903–914.

[18] Thepsonthi T, Özel T. An integrated toolpath and process parameter optimization for high-performance micro-milling process of Ti–6Al–4V titanium alloy. Int J Adv Manuf Technol 2014;75:57–75.

[19] Gelfi M, Attanasio A, Ceretti E, Garbellini A, Pola A. Micromilling of lamellar Ti6Al4V: cutting force analysis. Mater Manuf Processes 2016;31(7):919–925.

[20] Henry S, McAllister DV, Allen MG, Prausnitz MR. Microfabricated microneedles: a novel approach to transdermal drug delivery. J Pharm Sci 1998;87:922–925.

[21] Maton A, Hopkins J, McLaughlin CW, Johnson S, Warner MQ, LaHart D, Wright JD. Human Biology and Health. Englewood Cliffs, New Jersey, USA: Prentice Hall; 1993.

[22] Thepsonthi T, Milesi N, Özel T. Design and prototyping of micro-needle arrays for drug delivery using customized tool-based micro-milling process. Proceedings of the First International Conference on Design and Processes for Medical Devices, May 2–4, Brescia, Italy; 2012.

[23] Teresi LM, Lufkin RB, Reicher MA. Aysmptomatic degenerative disc disease and spondylosis of the cervical spine: MR imaging. Radiology 1987;164:83–88.

[24] Szpalski M. The mysteries of segmental instability. Bull Hosp Jt Dis 1996;55:147–148.

[25] Hilt ZJ, Peppas NA. Microfabricated drug delivery devices. Int J Pharm 2005;306:15–23.

[26] Kurtz SM, Devine JN. PEEK Biomaterials in Trauma, Orthopedic, and Spinal Implants. Biomaterials 2007;28(32):4845–4869.

[27] Ferguson SJ, Visser JM, Polikeit A. The long-term mechanical integrity of non-reinforced PEEK-OPTIMA polymer for demanding spinal applications: experimental and finite-element analysis. Eur Spine J 2006;15:149–56.

6

INKJET- AND EXTRUSION-BASED TECHNOLOGIES

KARLA MONROY, LIDIA SERENÓ, AND JOAQUIM DE CIURANA GAY

Department of Mechanical Engineering and Industrial Construction, University of Girona, Girona, Catalonia, Spain

PAULO JORGE BÁRTOLO

School of Mechanical, Aerospace and Civil Engineering, Manchester Institute of Biotechnology, University of Manchester, Manchester, UK

JORGE VICENTE LOPES DA SILVA

Technological Center Renato Archer, Centro de Tecnologia da Informação Renato Archer, Brazilian Ministry of Science and Technology, Campinas, São Paulo, Brazil

MARCO DOMINGOS

School of Mechanical, Aerospace and Civil Engineering, Manchester Institute of Biotechnology, University of Manchester, Manchester, UK

6.1 INTRODUCTION

Rapid prototyping, rapid manufacturing, solid, freeform fabrication, and 3D printing are well-known names for the production technologies that nowadays are grouped under the generic name additive manufacturing (AM), according to ASTM F2792-12a [1]. Additive manufacturing (AM) refers to a broad class of processes by which digital 3D design data are used to build a component in layers by continuous material deposition, layer-by-layer. The first AM processes were limited to the rapid

Biomedical Devices: Design, Prototyping, and Manufacturing, First Edition.
Edited by Tuğrul Özel, Paolo Jorge Bártolo, Elisabetta Ceretti, Joaquim De Ciurana Gay,
Ciro Angel Rodriguez, and Jorge Vicente Lopes Da Silva.
© 2017 John Wiley & Sons, Inc. Published 2017 by John Wiley & Sons, Inc.

production of visual prototypes (rapid prototyping) to understand the shape of a designed component or a system. However, additive manufacturing (AM) is no longer merely used as a production method for creating prototypes; today, it is more often used to create products with a high added value [2]. Due to the increasing interest and technological evolution, the improved dimensional accuracy achievable, and the superior properties acquired in the end materials, those production processes are nowadays also used for the production of single or small series of functional products [3].

There are more than 30 commercial AM processes with economic importance, each with its advantages and disadvantages and each appropriate for different materials and requirements. Such differences will determine factors such as the accuracy of the final part plus its material and mechanical properties. They will also determine factors such as how quickly the part can be made, how much postprocessing is required, the size of the AM machine used, and the overall cost of the machine and process [4, 15]. Typically, the AM technologies are grouped based on the type of raw materials or the method used to form those materials into finished products. In the approach to end the misuse of different marketing words to describe common processes, the ASTM International Committee F42 took on the task of developing a common terminology. Therefore, according to the ASTM definition, the available processes were classified into seven categories according to their baseline technology [5], as shown in Table 6.1 [6].

Binder jetting, which is commonly known as 3D printing, is an AM process in which a liquid bonding agent is selectively deposited to join powder materials. In this method, a modified inkjet head moves across a bed of powder spraying a binder material on the powder in the shape desired. It can bind plastic, metal, ceramic, and sand to form parts and molds.

Material jetting, also called "direct printing," is a process in which droplets of desired material is dispensed through a printhead. Originally evolved from systems

TABLE 6.1   Additive Manufacturing Categories as Classifies by ASTM

Category	Description
Binder jetting	Liquid bonding agent selectively deposited to join powder
Material jetting	Droplets of build material selectively deposited
Powder bed fusion	Thermal energy selectively fuses regions of powder bed
Directed energy deposition	Focused thermal energy melts materials as deposited
Sheet lamination	Sheets of material bonded together
Vat photopolymerization	Liquid photopolymer selectively cured by light activation
Material extrusion	Material selectively dispensed through nozzle or orifice

that used thermoplastics, the material jetting method has been modified to accept ceramic slurries or ceramic powders in wax or liquid binder carriers. Material jetting has significant challenges, including getting materials to flow through nozzles at reasonable speeds without clogging. Work is ongoing to improve the rheology of material systems for ceramic materials such as alumina, zirconium, and PZT. This method promises good surface finish and high tolerance for parts that can be printed and then fired to high density.

Rather than spraying on a binder, an alternative method to stick powder granules together is by applying heat as in powder bed fusion category. It originated with selective laser sintering (SLS), which uses a powder bed layer in a build box, similar to the binder jet method, but it is placed in a system that brings the powder to an elevated temperature and then exposes select areas to a laser beam. All powder bed fusion methods share certain characteristics, including one or more thermal sources for inducing fusion or sintering between particles, a method for prescribing fusion in a region of each layer, and mechanisms for adding and smoothing powder layers [6, 7]. At present, SLS printers can output objects using a wide range of powdered materials. These include wax, polystyrene, nylon, glass, ceramics, stainless steel, titanium, aluminum, and various alloys including cobalt chrome. Other closely related techniques are selective laser melting (SLM), selective heat sintering (SHS), and electron beam melting (EBM) [7].

As yet another variant of powder solidification, there is a technology called "directed energy deposition" (also known as "laser powder forming"). In directed energy deposition, focused thermal energy is used to fuse materials by melting as the material is being deposited [1, 7]. These processes work by heating and depositing material following a predefined path, layering on top of a platform, dropping material on top of previous layers to create the three-dimensional geometry. It can use wire or powder as materials and lasers or electron beams as the energy source. Typical materials for these processes are stainless steel, copper, nickel, cobalt, aluminum, or titanium.

In sheet lamination, as the name indicates, sheets of material are bonded together to form an object. The sheet lamination method was an early rapid prototype method used for plastics and paper that has now been adapted for ceramic tapes. It can bond paper using glue, plastic using glue or heat, and metal using welding or bolts. Extremely fine features with high tolerance can be obtained. This method is being used to form microfluidic devices with thousands of channels [6].

In the case of liquid, vat photopolymerization process solidifies thin layers together, using an ultraviolet (UV)-curable thermoset polymer liquid with a solid-state crystal laser to create the required geometry layer by layer, using computer-aided design (CAD) data. Vat polymerization is used for polymers but it can also be loaded with ceramic powders [6, 8]. Stereolithography (SL) is the most identified example of this type of AM process.

Finally, rather than solidifying a photopolymer, the last category is based on material extrusion. It is a process in which the material is selectively dispensed through a nozzle or orifice for a further bonding to itself in the next layer deposition [7]. Material extrusion includes fused deposition modeling (FDM), low-temperature deposition manufacturing (LDM), bioextrusion, and multiphase jet solidification (MJS). Most of the applications have focused on plastic systems, but the technique lends itself to ceramics. Examples of industrial ceramic parts include specialty crucibles and filters [6, 7].

This chapter is focused on the development of three of the seven additive approaches: material and binder jetting processes, which comprises the so-called "inkjet-based technologies" and material extrusion, which is fundamentally represented by the FDM process.

6.2 INKJET TECHNOLOGY

"Inkjet" is a collection of printing techniques that take small quantities of ink from a reservoir, convert them into drops, and transport the drops through the air to the printed medium (paper, transparencies, beverage containers, etc.) [9]. Inkjet printing is a type of computer printing that creates a digital image by propelling droplets of ink onto paper, plastic, or other substrates. From its initial use for product marking and date coding in the 1980s, and its development and widespread adoption for the desktop printing of text and images in the following two decades, inkjet technology is having an increasing impact on commercial printing for many applications. Its great flexibility permitted it to challenge the conventional methods for more specialized use, as for digital fabrication [10].

Now, the same exact processes by which individual drops of liquid are produced can be used to deposit materials to directly fabricate a part without intermediate tooling. In the additive process, the ink is replaced with thermoplastic and wax materials, which are held in a melted state. When printed, liquid drops of these materials instantly cool and solidify to form a layer of the part. Metals, ceramics, and polymers, with a wide range of functionality, can all be printed by inkjet methods, and innovative possibilities are also raised by the ability to print biological materials, including living cells [10, 11].

Inkjet technique, therefore, is one of the ultimate waves of technology development as a manufacturing process, since several features make it particularly attractive for manufacturing [10]:

- It is a digital process. The location of each droplet of ink or material being deposited can be predetermined on a two-dimensional grid. If necessary, the location can be changed in real time, to adjust for distortion or misalignment of the substrate or to ensure that a certain height of final deposit is achieved. Also, each product in a sequence can be easily made different from every other.

- It is a noncontact method; the only forces that are applied to the substrate result from the impact of very small liquid drops. Material can be deposited onto non-planar substrate (rough or textured), since the process can be operated with a stand-off distance of at least 1 mm between the printhead and substrate.

- A wide range of materials can be deposited. By selecting an appropriate printhead, liquids with viscosities from 1 to 50 MPa or higher can be printed. Several different methods can be used to generate printed structures. Multiple combinations of materials can be used, and inkjet printing can also be combined with other process steps, so that in principle complex heterogeneous and composite structures can be produced, with different materials distributed in all 3D.

- It is modular and scalable. Multiple printheads can be assembled to print in tandem, or one after the other to print different materials in sequence.
- In short, inkjet-based 3D printing is the only technology that can really simulate the true look, feel, and function of complex assembled goods. And this is thanks to the wide variety of materials available and the inkjet process itself—which can simultaneously jet different materials from the separate inkjet head nozzles [12]. This section explains the basic principles of the technology, its different approaches, a review of the range of materials, the various methods by which inkjet processes can be used, and main applications in the medical field.

6.2.1 Inkjet 3D Printing Technology

The additive fabrication technique of inkjet printing is based on the 2D printer technique of using a jet to deposit tiny drops of ink onto paper. In the additive process, the ink is replaced with thermoplastic and wax materials, which are held in a melted state. When printed, liquid drops of these materials instantly cool and solidify to form a layer of the part [11].

6.2.1.1 Basic Principle The inkjet printing process begins with the build material and support material being held in a melted state inside two heated reservoirs. These materials are each fed to an inkjet printhead that moves in the x-y plane and shoots tiny droplets to the required locations to form one layer of the part. Both build and support materials instantly cool and solidify. After a layer has been completed, a milling head moves across the layer to smooth the surface. The particles resulting from this are vacuumed away by a collector. The elevator then lowers the build platform and part so that the next layer can be built. This process is repeated for each layer until the part is completed. The part can then be removed and the wax support material can be melted away [13].

6.2.1.2 Approaches Inkjet printing technology comprises two main configurations [14]:

(a) binder jetting or bonding method
(b) material jetting or direct build-up method

Direct printing refers to processes where all of the part material is dispensed from a printhead, while binder printing refers to a broad class of processes where the binder or other additive is printed onto a powder bed that forms the bulk of the part [15].

6.2.1.3 Binder Jetting Binder jetting is a process by which the liquid bonding agent is selectively deposited through inkjet printhead nozzles to join powder materials in a powder bed. Binder jetting is similar to material jetting in its use of inkjet printing to dispense material. The difference is that, with binder jetting, the dispensed material is not build material but rather the liquid that is deposited onto a powder bed to hold the powder in the desired shape.

The binder jetting process was developed at MIT under the name of *powder bed and inkjet head 3D printing.* In 3D printing, binder jetting is a process in which layers of material are bonded to form an object. It is one of the best options for 3D printing in full color and has less noticeable layer definitions, making it an ideal choice for producing end-use products. The primary advantages of binder jetting are low cost, high speed, scalability, ease of building parts in multiple materials, and versatility for use with ceramic materials [15]. This process is also unique in fact it has been scaled up to print full architectural structures as big as a room. Objects created with the binder jetting process may not have the high-quality mechanical properties of other additive manufacturing techniques because of the materials used and the lower level of adherence between particulates [16, 17].

The process starts with a 3D printer software that reads an STL or other 3D file created in CAD software. The file is sliced creating layers of thin cross-sections of the object. These data are then sent to a printer where the object is printed by sticking together successive layers of a powder, sand, or metal material. The print starts by using an automated roller to spread the first layer of powder onto the build platform (Figure 6.1). Powder is fed from a piston to ensure that the layer is densely packed

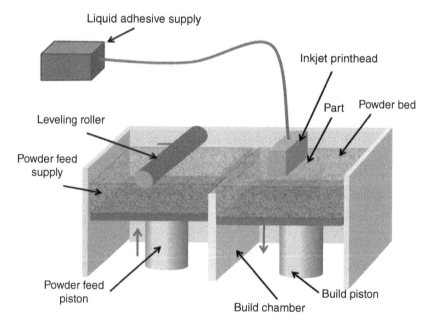

Figure 6.1   Binder jetting process overview. (*See color plate section for the color representation of this figure.*)

while excess powder is brushed to the sides. The printheads then apply the binder for that cross-section of the object. The process then repeats itself, followed by the next layer of material being spread onto the build platform. Supports are rarely required with this type of 3D printing as each horizontal slice is supported by the excess powder material below [16, 17]. When build cycle has ended, the loose powder is removed by vacuum, revealing the part. The part is porous and, if desired, can be infiltrated or fired in a postprocess step. This method is used for fine art ceramics, ceramic cores, and large industrial sand molds for metal casting [15, 18].

6.2.1.4 Material Jetting The material jetting is an approach that uses inkjet printing heads to deposit droplets of build material, which resembles traditional paper printers in more ways than many other 3D printing processes. The droplets are dispensed selectively as one or more printheads move across the build area. Two types of materials are predominantly used in this group of processes: wax and photopolymers. Material-jetting systems often use multi-nozzle printheads to increase build speed and to print different materials. Other advantages of this group of processes are as follows [19]:

- *Unique Material Properties.* Point-by-point controlled printing of multiple materials, allowing for unique material properties. Multi-material printing allows for an object with properties close to that of the desired final product. This capacity of using mixed materials eliminates the need to design and print each separate part in their respective materials and then assemble after printing, reducing the production time.
- *Postprocessing.* Easy to remove support material using either a water jet or sodium hydroxide solution. In addition, no postcuring is necessary.
- *Cleanliness.* Clean, compact process, ideal for use in an office-like setting.
- *Ease of Use.* Reliably autonomous process with minimal user interaction needed during printing.
- *Accuracy and Resolution.* Capable of fine details, quick printing, and high precision and accuracy, with a layer thickness that minimizes the "stair-stepping" effect of curved surfaces. These latter capabilities make it common for prototyping due to its high resolution (16 micron layer heights) and ability to match the look, feel, and function of the desired finished product.
- *Functional Assemblies.* Capable of constructing functional assemblies during the printing process, without the need for multiple builds, resulting in a high productivity output, or capacity for large prints.

In material jetting, the printhead moves around the print area jetting photopolymer as opposed to ink. The base materials are funneled into a dedicated liquid stream, which is then jetted onto the build tray to form the layers. One material is used to create support structures, while a second is used for the build material. Gel-like support materials are often used to support overhangs and complex geometries. UV lights surrounding the printhead pass over the material after it is jetted onto the build area and

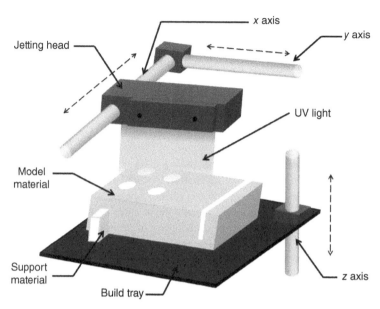

Figure 6.2 Material jetting working principle.

cure it, solidifying it in place. The curing process initiates a chemical polymerization reaction, which causes the plastic to dry and form a solid, fixing it in place. Repeating this process builds up the object one layer at a time (Figure 6.2). When the print is finished, the gel-like support material is easily removed by dissolving with water, usually accelerated by using a pressurized sprayer. The finished product is ready shortly after removing it from the machine, unlike other rapid prototyping processes, which can require lengthy postprocessing treatments [16, 17].

6.2.2   Materials in Inkjet-Based Technologies

Inkjet printing processes enable the processing of a wide range of polymeric, metallic, ceramic, and composite materials at room temperature and at a resolution of about 200 mm. For any application, appropriate selection of materials is first important as the primary consumable during inkjet printing. Therefore, the ink materials have to be uniformly dispersed in the solvent by additional suitable surfactant, in which no sedimentation is expected for the long run duration and the ink viscosity needs to be sufficiently low for jettability of droplet actuators as well. The sub-micrometer size of suspended particles, and in particular pigment and metal particles, is also limited to an extent that is determined by some thresholds (e.g., filter cavity and nozzle neck) inside the micro-liquid channels of printheads [20]. The surface tension also must be high enough and the pressure low enough, to hold the ink in the nozzle without dripping.

6.2.2.1   Polymers   In terms of raw materials that are placed into the inkjet printing system to create the 3D objects, there seem to be relatively few limitations on what can be used. Currently, plastics are the most widely used materials in additive manufacturing. Plastics are classified into two groups based on their behavior at high temperatures: thermoplastics and thermosetters. Thermoplastics retain their properties and can be repeatedly melted, cooled, and hardened. The build materials such as ABS (very common for 3D printing), PLA (available in soft and hard grades), PVA (dissolvable support material or for special applications), PC (requires high-temperature nozzle design and is in the proof-of-concept stage), and soft PLA (rubbery and flexible, available in limited colors and sources) are thermoplastics and are commonly used in binder jetting processes. Thermoset plastics, typically acrylic, acrylate, or epoxy materials, are used in material jetting systems. Material jetting also uses photopolymers, which can be combined together at different rates to create new materials with distinct tensile, flexural, and impact strengths, ranging from rigid to rubber-like and opaque to transparent [16, 17].

The prices of the polymers in these process are typically many times higher (even up to 125 times) than the equivalent materials in conventional manufacturing processes as in injection molding.

6.2.2.2 Metals

6.2.2.2 Metals In addition, the number of metal materials available is impressive, as there is an enormous interest in extending the range of material possibilities involving metals. Binder jetting systems produce metal parts comprised of stainless steel or tool steel with a bronze infiltrant. In the postbuild furnace cycle, the binder is burned out and bronze is infiltrated into the parts to produce metal alloys. Further material choices include many alloys such as aluminum, copper, titanium, gold, and silver. Similar to polymers, metal powders are more expensive compared with feedstock forms in conventional manufacturing.

6.2.2.3 Ceramics

6.2.2.3 Ceramics A wide range of materials have been developed for this technology, mostly by researchers. Printing into a metal powder bed was first demonstrated in the early 1990s, and concurrently, investigations into ceramic materials are also pursued. Traditional inkjet of ceramics involves the selective printing of a binder over a bed of ceramic powder. Alumina, silica, and titanium dioxide have been made with this process [21]. The fine powders needed for good powder bed density did not generally flow well enough to spread into defect-free layers. To counteract the problems encountered with recoating a dry powder bed, research on ceramic 3D printing has shifted to the use of a slurry-based working material. In this approach, layers are first deposited by inkjet printing a layer of slurry over the build area. After the slurry dries, the binder is selectively printed to define the part shape. This is repeated for each individual layer, at the cost of significantly increased build time. Recently, a variation of this method was developed to fabricate metal parts starting with metal-oxide powders. The ceramic is used until the furnace sintering step. While in the furnace, a hydrogen atmosphere is introduced, causing a reduction reaction to occur between the hydrogen and the oxygen atoms in the metal-oxide. The reduction reaction converts the oxide to metal. After reduction, the metal particles are sintered to form a metal part. This process has been demonstrated for several material systems, including iron, steel, and copper [15].

6.2.2.4 New Trend Materials   As mentioned earlier, there are few limits on what
materials can be used for inkjet manufacturing. Some of the unusual and recently
employed materials are as follows [22]:

- *Chocolate*. Material engineers have devised a way to use chocolate to obtain
  some innovative designs in this luscious material.
- *Bio-Ink*. This ink comprises stem cells and cells from a patient, which can be
  laid down, layer by layer to form a tissue. Human organs such as blood vessels,
  bladders, and kidney portions have been replicated using this technology. Using
  biocompatible binders such as water-based binders, these processes also allow
  the fabrication of constructs containing sensitive biomolecules such as cells,
  growth factors, or drugs.
- *Bone Material*. Washington State University printed a bone-like material com-
  prising silicon, calcium phosphate, and zinc. This bone-like material was integrated with a section of undeveloped human bone cells where later new bone
  grew successfully along the structure and the new material dissolved eventually
  without harming the patient.
- *Sandstone*. This material enables the production of 3D printed creations with
  almost any color. Fine designs for action figures, architecture, and character
  models are becoming highly popular with this material.
- *Glass*. Ground up glass powder is spread layer by layer, bonded with adhesive
  spray then baked resulting in 3D printed glass product.
- *Medication*. Engineers and doctors are working together to create 3D printed
  medication.
- *Skin*. Similar to bio-ink, printers can help in regenerative skin applications.

6.2.3   Inkjet Printing Methods

There are three types of inkjet methods: thermal phase, suspension, and UV curable
[11].

6.2.3.1 Thermal Phase Change Inkjet   During this process, the inkjet machines
hold the build and support materials at elevated temperatures in a reservoir until the
fabrication of the part begins. Once the process begins, the liquid material moves
through thermally insulated tubing to individual jetting heads. The jetting heads then
disperse the material in the form of tiny droplets to create the part geometry. This
so-called drop-on-demand (DOD) technology only places droplets where they are
required to form the part. The droplets begin to cool and harden immediately after
leaving the jetting head. When one layer is complete, the milling head passes over
to produce a uniform thickness. The material particles are vacuumed, the nozzles are
checked, and then the table is moved down for the next layer building. Once the part
is finished, the support structures are melted. This type of inkjet machine is capable of fast production of fine parts when using multiple heads; however, accuracy is
lowered.

6.2.3.2 Suspension Inkjet The main components of these inks are volatile organic compounds (VOCs)—organic chemical compounds that have high vapor pressure. Inkjet printing is performed by the deposition of ceramic suspensions, which dry by evaporation of the solvent that uses systems initially devoted to inkjet printing on paper. The high print speed of many solvent printers demands typically a special combination of heaters and blowers. The substrate is usually heated immediately before and after the printheads apply ink.

6.2.3.3 UV-Curable Inkjet These inks consist mainly of acrylic monomers with an initiator package. After printing, the ink is cured by exposure to strong UV light. The advantage of UV-curable inks is that they dry as soon as they are cured, they can be applied to a wide range of uncoated substrates, and they produce a very robust image. Disadvantages are that they require expensive curing modules in the printer, and the cured ink has a significant volume and results in a relief on the surface.

6.2.4 Inkjet Printing Systems: Processes and Machines

To fulfill the specific requirements of applications, basic inkjet printing processes should be comprehensively designed and realized with five major elements of system implementation: ink materials, substrate properties, droplet generation, electromechanical platforms, and printing algorithm [20]. In summary, the inkjet system is comprised of materials, machines, and the process itself. Materials, methods, and approaches have already been discussed; in this section, the common inkjet processes and commercial manufacturers that developed different inkjet printing devices are presented.

6.2.4.1 3D Printing Three-dimensional printing (3D printing) is a binder jetting approach invented at MIT and licensed to more than five companies for commercialization. 3D printing prints a binder into a powder bed to fabricate a part. Hence, in 3D printing, only a small portion of the part material is delivered through the printhead; most of the part material is comprised of powder in the powder bed. Typically, binder droplets (80 mm in diameter) form spherical agglomerates of binder liquid and powder particles as well as provide bonding to the previously printed layer. Once a layer is printed, the powder bed is lowered and a new layer of powder is spread onto it (typically via a counter-rotating rolling mechanism), very similar to the recoating methods used in powder bed fusion processes [15, 23].

Since the process can be economically scaled by simply increasing the number of printer nozzles, the process is considered a line-wise patterning process. Such embodiments typically have a high deposition speed at a relatively low cost (due to the lack of a high-powered energy source), which is the case for 3D printing machines. The printed part is typically left in the powder bed after its completion for the binder to fully set and for the green part to gain strength. Postprocessing involves removing the part from the powder bed, removing unbound powder via pressurized air, and infiltrating the part with an infiltrant to make it stronger and possibly to impart other mechanical properties.

The 3D printing process shares many of the same advantages of powder bed processes. As the parts are self-supporting in the powder bed, support structures are not needed. Similar to other processes, parts can be arrayed in one layer and stacked in the powder bed to greatly increase the number of parts that can be built at one time. Finally, assemblies of parts and kinematic joints can be fabricated since loose powder can be removed between the parts [15].

MIT licensed the 3D printing technology according to the type of material and application that each licensee was allowed to exploit. Z-Corporation, Inc. is one company that markets machines that build concept models in starch and plaster powder using low viscosity glue as binder. As of January 2008, Z-Corporation markets three models of printers: the ZPrinter 310 Plus, the ZPrinter 450, and the Spectrum Z510. At the other end of the spectrum, ExOne Company markets line machines such as the R1, R2, and S15, in which the first two models include applications in prototypes of metal parts and some low-volume manufacturing, as well as tooling. The model S15 is intended for companies with large demands for castings, such as the automotive, truck, and heavy equipment industries. The S15 can build metal parts and sand casting molds and cores in foundry sand metal powder.

Smaller manufacturers exist as Microjet Technology Co., Ltd and Voxeljet Technology GmbH. Finally, personal 3D printers can be acquired by open source projects such as RepRap, MakerBot, Solidoodle, Printrbot, Ultimaker, Aleph Objects, and Kickstarter.

6.2.4.2 Multi-Jet Modeling (MJM) Multi-jet modeling, also known as thermojet, is a rapid manufacturing process used for concept modeling. The system generates wax-like plastics models, which are less accurate than stereolithography. The MJM machines use a wide area head with multiple spray nozzles. These jetting heads spray tiny droplets of melted liquid material that cool and harden on impact to form the solid object [23]. The printer creates the model one layer at a time by spreading a layer of resins and printing a binder in the cross section of the part using an inkjet-like process. The technology uses one material for supports and another for parts, is relatively slow, and includes a milling operation between layers. This is repeated until every layer has been printed. The strength of bonded powder prints can be enhanced with wax or thermoset polymer impregnation. Supports are used to support overhanging features during construction. Supports are removable or dissolvable upon completion of the print [24]. The process is commonly used for creating casting patterns for jewelry, dental, medical, and other precision casting applications.

MJM is a product of 3D systems from the makers of the SLA system. Systems such as the 3DSYSTEMS ProJet series spray photopolymer materials onto a build tray in ultrathin layers until the part is completed. Each photopolymer layer is cured with UV light or solidified by temperature after it is jetted. The gel-like support material, which is designed to support complicated geometries, can be removed by hand or hot air drying [24]. As a multiple nozzle system is fast compared with most other AM techniques, it produces good appearance models with minimal operator effort. The main market that this system is targeted at the engineering office where the system must be nontoxic, quiet, small, and with minimal odor [25].

Solidscape, Inc. also manufactures and sells machines to produce wax patterns for metal castings by MJM. The Solidscape machines employ the material jetting process to produce parts with a high level of precision, accuracy, and surface finish. In 2012, it launched the first of its 3Z line, which includes four models; each features a touch screen, automatic status monitoring and fault detection, and an auto-calibrating printhead. The 3Z line has a resolution of 5000 dpi in the x-y plane, 8000 dpi in the z direction, and an accuracy of ± 0.001 mm/mm in all directions.

6.2.4.3 PolyJet Process The PolyJet process uses a high resolution inkjet technology combined with UV-curable materials to quickly and economically produce highly detailed and accurate physical prototypes. The technology works by depositing very small droplets through two or more jetting heads (one set for the model and one set for the support material) that spray outlines of the part, layer by layer. It uses layers as thin as 16 microns (0.0006 in) making it an excellent option for presentation models and master patterns for advanced urethane castings. The liquids used are photopolymers, which get cured nearly instantly by a UV lamp within the printer, creating a solid, plastic-like model that is precise and accurate. The support material is a gel-like substance, which is easily washed away. Then, a UV bulb cures the droplets, and the printhead covers the build area to make a very thin layer of the prototype part. The process is repeated until the prototype is built.

The final models have a smooth finish and are ready for sanding, painting, drilling, or tapping. PolyJet parts are produced without any support vestiges, and with excellent up-facing and down-facing surfaces. The part fits within a 5 in × 5 in × 5 in (127 mm × 127 mm × 127 mm) build volume; other AM technologies may be more cost-effective for larger parts.

It is best suited for applications where accuracy, detail, and surface finish are important. Typical applications include electronic components and connectors, electronic packaging, presentation models, knobs, buttons, medical devices, fittings, valves, and parts with complex interior features. PolyJet has also been used for tooling in low-temperature applications. It makes an excellent mold with releasable surfaces without requiring a secondary process for noncosmetic applications [26].

Objet introduced the first 3D printer that combines ink jetting with photopolymers. The result is a high resolution additive manufacturing process that has the unique ability to print parts with multiple materials. Objet introduced the multi-material printing in 2007, which was acquired by jetting two different build materials in different proportions; the systems can produce parts, or regions of parts, with up to 14 different material compositions [27]. Stratasys Ltd. acquired Solidscape in 2011, so it is wholly operated by its owned subsidiary. Objet has introduced the Objet1000 model; the machine which owns a building envelope 10 times larger than its predecessors Connex500 and Eden 500V. The Objet1000 features digital materials, also known as Connex technology. Later, Objet launched the Objet30 Pro, capable of printing seven different materials, and in 2013, a derivative model Objet30 Orthodesk was launched as a new system particularly for dental materials. The price of both systems is around $32,000. Smaller manufacturers such as Keyence Corp also offer Polyjet systems.

6.2.4.4 Printoptical Technology The process prints structures using modified wide format industrial document inkjet printing equipment. Transparent droplets of a UV-curable polymer are jetted and then cured by strong UV lamps that are integrated into the printhead. The results of the process are freeform shapes that may include transparent prisms or lens, as well as full-color 3D graphics and textures. The piezoelectric controllable printhead provides a resolution of 1440 dpi or greater.

As the material is deposited in discrete drops, the resulting surface is smooth. This is accomplished by delaying the time between the jetting of the droplets and the application of UV light, which gives the polymer time to flow and for each droplet to lose its spherical form. Optical quality is achieved with no postprocessing [28].

LuXeXcel Group B.V. offers services based on printoptical. The current commercial applications offered are in lighting, solar, window foils, treatments, advertising graphics, and digital art. It offers 3D printing services, but does not sell systems yet. The company optimized the process in 2012 by introducing tinted polymer inks and improved the precision of the hardware and optical clarity. According to LuXeXceL, printoptical technology avoids the complicated and costly conventional processes, such as tooling, injection molding, diamond turning, polishing, and grinding which are typically used to produce many types of optical parts.

6.2.4.5 Bioprinting Three-dimensional bioprinting is the process of generating spatially controlled cell patterns in 3D, where cell function and viability are preserved within the printed construct. Using 3D bioprinting for fabricating biological constructs typically involves dispensing cells on a biocompatible scaffold using a successive layer-by-layer approach to generate tissue-like three-dimensional structures. Given that every tissue in the body is naturally compartmentalized of different cell types, many technologies for printing these cells vary in their ability to ensure stability and viability of the cells during the manufacturing process.

Three-dimensional printed human tissue is created by using modified printer cartridges and extracted cells, sourced from patient biopsies with respect to examining cancer cells or stem cells. They are grown using standard techniques and cultured in growth media in dishes, allowing them to multiply. Once enough cells have grown, they are collected and formed into spheroids or other shapes and loaded into a cartridge to create bio-ink. The bio-ink is loaded into a bioprinter along with a cartridge of hydrogel, a kind of synthetic matrix effectively used as a kind of scaffolding for building 3D layers of cells. The printer prints a layer of the water-based gel, followed by a layer of bio-ink cells, and so on. The layered cells naturally fuse together as the layers are built upon. Bioinert hydrogel components may be utilized as supports, as tissues are built up vertically to achieve three-dimensionality or as fillers to create channels or void spaces within tissues to mimic features of native tissue [29].

Once the desired amount of layers is printed, the printed tissue is left to mature and grow as a structure, during which time the hydrogel is removed. Other researchers experimenting with bioprinting have used a sugar and water solution as a form of support for the vascular structures to great success. Currently, printed tissues are generally used for medical research—introducing disease to monitor how the tissue reacts and how future treatments may be developed. In the future, it is very likely that 3D printers will be used to create simple tissues for implanting into current organs and partial organs. The printing of whole organs, if approved, could be a reality within the next decade [30].

An early-stage medical laboratory and research company, called Organovo, designs and develops functional, three-dimensional human tissue for medical research and therapeutic applications. In 2010, the company printed the first human blood vessel without the use of scaffolds. Recently, Organovo bioprinted its first 3D liver tissue with the NovoGen MMX Bioprinter for testing purposes and can create 24 strips of liver tissues within a single plate [31].

6.2.5 Medical Applications of Inkjet Technology

Inkjet printing offers the advantages of excellent accuracy and surface finishes. However, the limitations include slow build speeds, few material options, and fragile parts. As a result, the most common application of inkjet printing is prototypes for form and fit testing. Other applications include jewelry, medical devices, and high precision products. This section focuses on some of the activities of inkjet-based technology in the field of medical treatment and health care, as this technology is capable of delivering personal solutions based on the need for custom-made products for treatments. Today, the application in health care is one of the most promising and is one of the most research-intensive production application areas due to its economic potential.

6.2.5.1 Biomodels and Prosthetics As mentioned previously, each of our body varies in its geometry and, therefore, medical applications are driven by patient's unique requirements of shape, functionality, and cost. The fulfillment of customization that the inkjet technology can reach is a very good solution for specific patient needs. The well-known use is the medical model (biomodel) for precise surgical planning. A biomodel is a perfect replica of the patient, produced based on computerized tomography (CT) or the magnetic resonance imaging (MRI) data set of this patient. The data set is processed using specialized software that can generate a virtual 3D model in a specific file format with the anatomical region of interest. During this process, a simulation of the interaction between the implant and the anatomy structure could be useful. Biomodels can model craniofacial bones, orthodontic, jaw, and general bone tissue. Many models may benefit from having different colors to highlight important features. Such models can display tumors, cavities, vascular tracks, etc. [32].

Initially, the technology was capable of creating biomodels that looked anatomically correct, but as the technology improved it is now possible to use them in combination for fabrication of close-fitting prosthetic devices. Many prosthetics are comprised of components that have a range of sizes to fit a standard population distribution. Greater comfort and performance can be achieved where some of the components are customized, based on actual patient data [15]. Examples can be maxillofacial, hip joints, and joint articulations.

6.2.5.2 BioMEMS MEMS devices that are implantable in humans have both great potential and significant challenges. Continuous monitoring and adjustment of biological functions has great potential for more effective monitoring and treatment regimens. However, implantable MEMS devices must deal with biocompatibility and biofouling issues [33]. The electrochemical measurement sites on a ceramic substrate of an implantable microelectrode can be coated with a glutamate oxidase enzyme and overcoated with glutaraldehyde, a fixative, both using inkjet printing. Currently, this type of probe is used for research into neurological function and diseases such as Parkinson. Future variants of this type of device could be used as an indwelling sensor in humans, which would monitor and report brain function as part of an overall treatment plan, similar to blood glucose measurement and insulin injection [34].

6.2.5.3 Drug-Eluting Devices Delivery of drugs by implantable devices is of as much interest as are implantable sensors; the current research includes drug delivery for diabetes, cancers, cardiac disease, and neurological disorders. The most widely used implantable drug delivery device today is the drug-eluting stent, used to keep coronary arteries open after an angioplasty procedure. Stents consist of a cylindrical scaffold, which is a largely open structure, composed of a rectangular wire mesh defined by interconnected struts. Typical strut widths are in the 50–100 µm range, so overspray and resulting wastage, though reduced, is still an issue. The complex structure of a metal stent must allow the device to collapse to travel through the blood vessels, then deploy, and lock at the desired location. To prevent tissue from growing over the stent and blocking the artery again, drugs are embedded into the stent. This requires placing the drug onto one side of complex features with 50–150 µm in width [10]. The use of drug-eluting stents in the field of interventional cardiology has been extremely successful in reducing the incidence of restenosis from 20% to 30% [35].

A number of companies are using inkjet for this process, as it offers unique advantages for the coating of very small medical devices with drugs and drug-coating combinations, especially in cases where the active pharmaceutical agent is very expensive to produce and wastage is to be minimized. For medical devices such as drug-containing stents, the advantages of inkjet technology result from the controllable and reproducible nature of the droplets in the jet stream and the ability to direct the stream to exact locations on the device surface [34].

6.2.5.4 **In Vivo** *Devices: Scaffolds* *In vivo* devices are artificial devices put into the human body or living organisms. Examples include scaffolds, pins, plates, rods, screws, valves, sutures, grafts, and fixations, which are manufactured to suit a variety of medical situations. Scaffolds are permanent or temporary porous structures implanted to favor tissue regeneration. Scaffolds aim to serve as a framework to structurally reinforce the defect and prevent the collapse of surrounding tissue, to orchestrate endogenous host cell response, including major cellular processes (i.e., migration, proliferation, and differentiation), and to degrade in a timely manner, in order to induce tissue ingrowth and proper orientation of the extracellular matrix (ECM) [36, 37].

A scaffold designed for bone, cartilage, and OC regeneration must meet several key parameters: biocompatibility, biodegradability, bioactivity, appropriate mechanical properties, and appropriate pore and surface characteristics [38]. A variety of scaffold fabrication techniques are available and can be classified into two categories: conventional and AM techniques, in which inkjet-based technologies are used [39].

Toh et al. [37] investigated the production of scaffolds using 3D printing/plotting with a starch-based polymer. A blend of starch-based polymer powders suitable for 3D printing (50% cornstarch, 30% dextran, and 20% gelatin) was developed. The structures were postprocessed by infiltrating with a co-polymer to increase the strength and water absorption resistance. The scaffolds were examined for their microstructure, porosity, mechanical properties, and water absorption levels. The results showed that it was possible to produce scaffolds with these materials using 3D printing, but its biocompatibility was not determined.

The printing process was also able to meet the desirable mechanical properties for *in vivo* applications in the generation of an osteochondral scaffold. This bonding method produced structures consisting of a cloverleaf shaped portion made of an osteoconductive PLGA/TCP composite that was 50% porous and a PLGA/PLA upper region with 90% porosity for seeding of chondrocytes. The transition region was designed to have a gradient of porosity (85%, 75%, and 65%) to ensure integration of the two distinct scaffold regions and to prevent delamination [40].

Cui et al. [41] used a material jetting method with simultaneous photopolymerization of poly(ethylene glycol)dimethacrylate (PEGDMA) hydrogel with human chondrocytes to repair defects in OC plugs (3D biopaper) in layer-by-layer assembly. The resulting printed PEGDMA was close to the requirements for human articular cartilage, and chondrocytes maintained the initially deposited positions (resembling zonal structure); and its viability increased 26% due to simultaneous photopolymerization.

Boland et al. [42] used an inkjet printing system to print 0.25 M calcium chloride ($CaCl_2$) onto liquid alginate/gelatin solutions. Similarly, Cohen et al. developed a robotic system to deposit alginate hydrogels seeded with articular chondrocytes. The alginate-cell suspension at a concentration of 33–106 cells/ml was cross-linked with calcium sulfate. After printing, 94.5% of cell viability was observed.

Finally, Sachlos et al. used an indirect approach to produce collagen scaffolds with complex internal morphology and macroscopic shape by using a 3D printing sacrificial mold. A dispersion of collagen was cast into the mold and frozen. The mold was then dissolved with ethanol and the collagen scaffold was critical point dried with liquid carbon dioxide. Other research works, such as those of Limpanuphap and Derby and Park et al., have also exploited the capabilities of 3D printing for tissue engineering [42].

6.2.5.5 Tissue and Organ Printing Living tissues are composed of many cell types that are all arranged in a very specific order in three-dimensional space. Maintaining this order is essential to ensure that engineered tissues and organs maintain the same functionality of the original body parts. Additive processes, such as 3D printing, are now able to produce organs and complex tissues since droplet-based printing technology can arrange cells and macrotissues (tissue spheroids) in predetermined locations with high precision [43, 44]. This methodology could eventually lead to the fabrication of complex, multicellular soft tissue structures such as liver, kidneys, and even heart.

Researchers are using modified inkjet technology to build a variety of tissue and organ prototypes [44]. Two types of inkjet systems have been used: thermal and piezoelectric inkjet printers. Cui and Boland reported the fabrication of functional human microvasculature networks by printing fibrin gel and human microvascular endothelial cells, using a modified thermal inkjet printer. It was possible to observe the cells aligned inside the fibrin channels forming a tubular structure. The produced microvasculature presented good integrity after 21 days of culture. Piezoelectric inkjet printers have also been used to print living human fibroblast cells and bovine vascular endothelial cells. However, piezoelectric systems are characterized by small-diameter jetting heads, resulting in small material droplets and lower cell concentrations (104–106 cells/ml). Several researchers used pressure-operated printing systems to print living cells and cell aggregates [42].

A material jetting approach can emit a stream of hydrogel microparticles to an exact coordinate offering another possible method for organ printing. This process is based on the deposition of bio-ink particles in well-defined topological patterns into bio-paper sheets of biocompatible gels. After the deposition, the construct is transferred to a bioreactor to fuse the bio-ink particles. Three-dimensional myocardial patches can be formed using this approach [42].

6.2.5.6   In situ Biofabrication for Skin Repair and Regeneration   *In situ* biofabrication involves the fabrication of substitutes for tissue repair and regeneration directly in the lesion of the patient. In this case, biofabrication systems are combined with real-time imaging techniques and path-planning devices, enabling the controlled deposition of biomaterials with or without encapsulated cells into the lesion site. *In situ* biofabrication has great potential for clinical applications, due to its minimally invasive nature, the possibility of eliminating the need for postprocessing operations, the ability to fabricate patient-specific biological substitutes, and reduced intervention time. Theoretically, this new concept can be applied for the regeneration of different tissues, but recent works are only focused on osteochondral, bone, and skin defects [45].

The current strategies for skin regeneration still present major pitfalls such as inadequate vascularization; poor adherence to the wound bed; inefficient elasticity; inability to reproduce hair follicles, sweat, sebaceous glands, and pigmentation; and the possibility of rejection and high manufacturing costs. Some of these limitations could be addressed by a printed skin substitute, which can cover the wound to prevent contamination, provide adequate moisture to avoid dehydration, give immediate and effective relief to patients, and can induce skin regeneration, due to its functional composition and cell combination [45].

Langnau [46] developed a portable inkjet delivery system for the *in situ* printing of skin cells into the lesion site. This system of skin regeneration was evaluated through the printing of human keratinocytes and fibroblasts into a full-thickness skin lesion, and results showed the complete closure of the wound after 3 weeks, as well as the formation of skin with properties similar to that of healthy skin. Histological analysis revealed that the new skin contained organized dermal collagen and a fully formed epidermis.

6.3 MATERIAL EXTRUSION TECHNOLOGY

Material extrusion is the most common and simplest 3D printing technique. It is suitable for domestic use and can be deployed in almost every environment. It was invented by S. Scott Crump, founder of Stratasys Ltd., patented filed (US Patent 5,121,329) in 1989 and awarded in 1992. The extrusion process consists of simultaneously advancing and melting of a filament of material through a computer-controlled nozzle. The material then flows through the nozzle under pressure, which must fully solidify while remaining in that shape. Deposited layers are fused together as the melted material quickly solidifies to form a solid three-dimensional object. If the pressure remains constant, then the resulting extruded material will flow at a constant rate and will remain a constant cross-sectional diameter. This diameter will remain constant if the travel of the nozzle across a depositing surface is maintained at a constant speed that corresponds to the flow rate [46]. The precision and quality of the final result depends mostly on the diameter of the printing head and on the thickness of the filament.

6.3.1 Material Extrusion—General Principles

The operation principle of the technology involves the extrusion of highly viscous materials through a nozzle. However, the extrusion process can be better understood, if described by its key steps and features that a common extrusion-based system possess, which are as follows [15]:

(a) material loading
(b) material liquefaction
(c) pressure application for material flow through the nozzle
(d) extrusion
(e) plotting
(f) material bonding
(g) support structures

These characteristics are discussed in separate sections, for further description in detail and for a better understanding of the complexities of the extrusion-based 3D manufacturing.

Material Loading In the extrusion system, a chamber from which the material is extruded is very essential. This chamber is preloaded with bulk material, which is typically supplied as solid in pellets or powder form, although it is more useful as a filament for continuous supply. The chamber itself is the main location for the liquefaction process. Depending on the form of the material, there are three different approaches of material feeding. Figure 6.3 shows an overview of the principles, which are up to now not commonly available: (1) filament, (2) syringe, and (3) screw based [2].

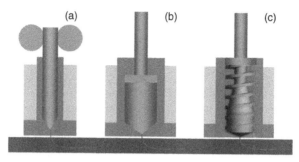

Figure 6.3 Overview based material extrusion principles (a) filament-based extrusion, (b) syringe-based extrusion, (c) screw based.

- *Filament-Based Extrusion* In the popular commercially available filament-based extrusion process, a spooled filament is fed into the liquefier using a pinch feed mechanism [15]. The incoming solid filament acts as a plunger to push and extrude the liquefied material through the nozzle [47]. This extruded material is deposited according to the pattern generated from the 3D model on the build platform or the previous layer [2].

  In comparison to conventional extrusion, the filament needs to be produced in a very tight diametric tolerance, which cannot be achieved in conventional extrusion processes [48, 49]. Since the filament-based 3D printing machine drive pushes the filament feedstock through the liquefier, a filament with a bigger diameter will block the system (Figure 6.4a), and a filament with a smaller diameter will not touch the wall of the extruder and will cause material to rise between the wall and filament (Figure 6.4b). Further drawbacks of this approach are buckling [47, 50, 51] (Figure 6.4c) and slippage of the filament on the pinch wheel, causing an interruption on the building process.

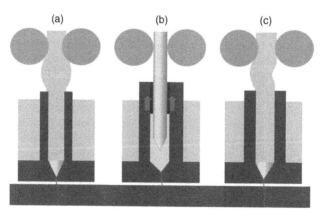

Figure 6.4 Filament extrusion: problems caused by the (a) filament, (b) improper diameter filament, and (c) buckling.

- *Syringe-Based Extrusion* To overcome the limitations of the filament-based extrusion process, especially in terms of amount of feed materials used and the range of the materials that can be processed, syringe-based extrusion process arose (Figure 6.3b). This configuration typically uses materials that solidify due to a chemical solidification, commonly used for biochemical applications where biocompatibility is required [15]. A reservoir is filled with the material and heated to the processing temperature. A force or displacement controlled plunger pushes the material out of the reservoir according to the generated path from the 3D model. If the material is in a liquid form, then the ideal approach is to pump this material through a syringe.

  Despite the satisfying results of the system, there are some disadvantages of syringe-based printing. The material in the syringe can end in thermal degradation due to the amount of time at elevated temperature, resulting in poor material properties. In addition, differences in melt viscosity can be found in the process caused by an inhomogeneous temperature distribution inside the barrel. In addition, during the production of large objects, the syringe needs to be refilled repetitively causing interruptions in the building process and air encapsulations; and the cooling down of the partial printed product and syringe can result in poor adhesion between the influenced layers [52].

- *Screw-Based Extrusion* As a solution for the problems in filament- and syringe-based extrusion, especially the limited range of available materials, a screw-based extrusion process was designed. Commercially available granules can be fed directly into the barrel; a continuous process can be realized and polymers suffer less thermal degradation.

  Materials that are fed through the system under gravity require a plunger or compressed gas to force it through the narrow nozzle. Pellets, granules, or powders can now be fed through the chamber with the aid of a screw. Screw feeding not only pushes the material to the base of the reservoir but can be sufficient to generate the pressure needed to push it through the nozzle as well [15]. Granules, which are the typical shape for thermoplastic bulk material, are fed into the hopper and transported to the nozzle by a screw. Heat is applied to melt the granules. Pressure build-up by the screw geometry is needed to expel trapped air between the granules and to overcome the backpressure induced by the shape of the nozzle, acting as the die for the extrusion process.

  The construction of this system requires a three-section-screw to avoid trapped air in the material melt, and to prevent interruptions during the extrusion; besides the screw, a barrel, and a nozzle also frame the mounting of the printhead, and heaters can be added to assist the melting process of the material (Figure 6.3c).

Material Liquefaction The extrusion method works on the principle that if a material can be presented in a liquid form that can quickly solidify, then it is suitable for this technique. This means that the material held in the chamber will become a liquid that eventually can be pushed through the die or the nozzle (Figure 6.5).

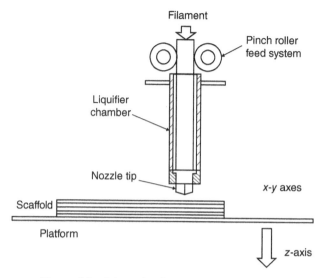

Figure 6.5   Schematic of extrusion-based systems.

As mentioned earlier, this material can be in the form of a solution; however, it is more common that the material is melted. The change in the state of the material can be done by two primary approaches. The first and most commonly used is by the temperature, wherein the molten material is liquefied so that it can flow out through the nozzle. Such heat would normally be applied by heater coils wrapped around the chamber and ideally this heat should be applied to maintain a constant temperature in the melt. In this approach, special care should be taken in the temperature peaks and duration as it can degrade the materials, and higher temperatures also require a cooling phase after extrusion.

An alternative approach is to cause solidification by a chemical change that permits bonding to occur, either by a curing agent, residual solvent, reaction with air, or simply by the drying of a "wet" material. As a result, parts become fully stable after curing or dry out. This approach is more appropriate in biochemical applications where materials must have biocompatibility with living cells, and therefore, choice of material is very restricted. However, industrial applications may also exist [15].

Extrusion The extrusion is primarily done through a nozzle, which determines the shape and size of the filament. The diameter of the nozzle also determines the minimum feature size that can be created. A larger nozzle diameter will enable material to flow more rapidly but would degrade the geometrical precision. No feature can be smaller than the nozzle diameter, and in practice, details should be approximated to the nozzle diameter to realistically replicate them with success. Therefore, it is a process more suitable for features and wall thicknesses of at least two times the nominal diameter of the extrusion nozzle used.

The material flow through the nozzle during the extrusion is controlled by the pressure drop between the chamber and the surrounding atmosphere [15].

Solidification Ideally after extrusion, the material should remain the same in shape and size. However, the material may change its shape due to gravity and surface tension, while size changes can be owed to cooling and drying effects.

If the material is extruded in the form of a gel, the material may shrink and become porous. If the material is extruded in a molten state, it may also shrink when cooling. If this cooling has a significant nonlinear effect, the resulting part will distort after cooling. By ensuring a minimum temperature differential between the chamber and the atmosphere and the control of a gradual and slow cooling process, later alterations can be minimized [15].

Plotting and Path Control Extrusion-based systems use an extrusion head, which is typically carried on a plotting system that allows movement in the horizontal plane and a platform that does indexing in the vertical direction to allow formation of individual layers. The plotting system must be coordinated with the extrusion rate to guarantee a smooth and consistent deposition. As the mechanical extrusion head needs to be moved in the horizontal plane, then the most appropriate mechanism is a standard planar plotting system. Cheap systems often make use of belts driven by stepper motors, while higher cost systems typically use servo drives with lead screw technology. This system is usually controlled by a computer-aided manufacturing (CAM) software package running on a microcontroller. This precise control depends on a significant number of parameters, which are described as follows [15]:

- *Input Pressure*. The force applied to the material is changed repeatedly during a build, which results in a corresponding output flow rate change.
- *Temperature*. Ideally, a constant temperature within the melt should be achieved; however, this is unavoidable and changes in the flow characteristics will occur. Therefore, temperature control should be carried out and compensate the thermal variations by the input pressure. As the heat builds up, the pressure should drop slightly to maintain the same flow rate. Furthermore, during the cooling phase, the different geometries will cool at different rates affecting the part properties; for that reason, the temperature should be taken care of till the end of the process.
- *Nozzle Diameter*. During the building process, this remains unchanged, although the systems allow interchangeable nozzles that can be used to compensate speed in favor of precision.
- *Material Characteristics*. Information of material properties such as viscosity should be taken into account, as it can assist in the understanding and control of the material flow through the nozzle, which is difficult to predict.
- *Gravity and Other Factors*. If no pressure is applied to the chamber, it is likely that the material will still flow, as the mass of the molten material within the chamber causes a pressure head. Surface tension of the melt and drag forces at the internal surfaces of the nozzle may retard this effect.

The factors stated earlier should be taken into consideration in order to have a better control of the material flow from the nozzle and the corresponding precision of the final part.

Bonding Once the material has been extruded, it must solidify and bond with adjacent material. For heat-based systems, there must be sufficient residual heat energy to activate the surfaces of the adjacent regions, while gel-based systems must contain residual solvent or a wetting agent in the extruded filament to ensure that the new material bonds with the adjacent regions. In both cases, if there is insufficient energy, the regions may adhere, but then the materials can be easily separated and a fracture surface can happen. In contrast, if too much energy occurs, the previously deposited material can flow and result in a poorly defined part.

Support Generation As in AM, complex features can be achieved but such features sometimes must be kept in place by the additional fabrication of supports, which can take two common forms:

• similar material supports
• secondary material supports.

If the extrusion-based system has only one chamber then supports must be made using the same material as the part. This may require supports designed and placed carefully in order to be easily separated at the end. One possible method is to increment the layer separation distance when depositing the part material on top of the support material or vice versa; this can affect the energy transfer to result in a fracture.

However, the most effective way to remove supports from the part is to fabricate them in a different material. In order to achieve this, the equipment should have a second extruder that can work in parallel. The disparity in material properties can be used to make the supports easily distinguishable from part material, visually, mechanically, or chemically.

6.3.2   Extrusion-Based Technologies

6.3.2.1   Fused Deposition Modeling (FDM) The most commonly known extrusion-based technique is the FDM. This process was named, invented, and patented by Scott Crump, owner of Stratasys in 1988 [53]. Other manufacturers refer to the same process as "thermoplastic extrusion," "plastic jet printing" (PJP), the "fused filament method" (FFM), or "fused filament fabrication" (FFF) [7]. FDM is a clean, simple-to-use, office friendly 3D printing process. The process is capable of creating functional prototypes, tooling, and manufactured goods from engineering thermoplastics, as well as medical versions of these plastics. It is possible to produce complex geometries and cavities that would be difficult to build with traditional manufacturing methods [54]. It usually has a resolution ranging from hundred to several hundred micrometers, and unlike some other additive fabrication processes, it requires no special facilities or ventilation and involves no harmful chemicals and by-products [55]. The major strength of FDM is in the range of materials and the effective mechanical properties of resulting parts made using this technology. Parts made using FDM are amongst the strongest for any polymer-based additive manufacturing process.

The FDM process for rapid prototyping integrates three key components of the system: software, hardware, and materials, which are explained later.

6.3.2.2 Generating a Part with FDM FDM builds three-dimensional parts by melting and advancing a fine ribbon of plastic through a computer-controlled extrusion head, producing parts that are ready to use. This principle and its system can be seen in Figure 6.6.

FDM can be organized as a quick three-step process [56]:

1. *Preprocessing* The model is imported into a software program, which mathematically slices the conceptual model into horizontal layers, and deposition paths are created as an STL file. The preprocessing software calculates sections and "slices" the part design into many layers, ranging from 0.005 in. (0.127 mm) to 0.013 in. (0.3302 mm) in height. The software then uses this information to generate the process plan that controls the FDM machine's hardware.

2. *Production* In FDM, the model or part is produced by extruding small beads of material, which harden immediately to form layers. A thermoplastic filament or metal wire that is wound on a coil is unreeled to supply material to an extrusion nozzle head. The nozzle head heats the material and turns the flow on and off. The building process starts, where two materials, one to make the part and one to support it, enter the extrusion head. Inside the extrusion head, the filament is melted into liquid (1° above the melting temperature) by a resistant heater. The head traces an exact outline of each cross-section layer of the part. As the head

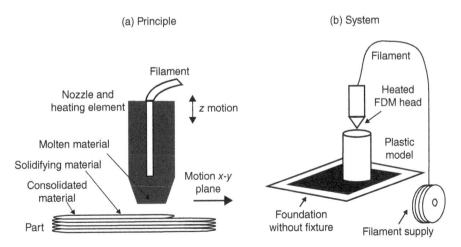

Figure 6.6 Fused deposition modeling process.

moves horizontally in the x and y axes the thermoplastic material is extruded out a nozzle by a precision pump. The material solidifies in 1/10 s as it is directed on to the workplace. After one layer is finished, the extrusion head moves up a programmed distance in z direction for building the next layer. Each layer is bonded to the previous layer through thermal heating. The overall tolerance of the process is +0.005 in.

3. *Postprocessing* When the part is complete, the support material is removed by either washing or stripping it away. Also, it has to be taken into account that many parts coming off these machines will not be immediately suitable for the final application and that there may be an amount of finishing required. To assist in this, a range of finishing stations can be designed to be compatible with various FDM materials. Finishing can be a mixture of chemically induced smoothing (using solvents that slightly melt the part surface) or burnishing using sodium bicarbonate as a light abrasive cleaning compound [15].

6.3.2.3 FDM Materials In relation to materials, FDM technology allows a variety of modeling materials and colors for model building. FDM uses two materials to execute a print job: modeling material, which constitutes the finished piece, and support material, which acts as scaffolding. Materials used in FDM must satisfy the requirements of the designer, their subsequent application, and their integration into the process. An FDM material must also have an adequate flexural modulus and strength to be formed into a filament, spooled, and used as a piston to pump the material through the head, liquefier, and tip. In addition, it must have sufficiently low viscosity to be pumped through the same hardware and also produce well-defined road widths over a broad range of geometries and deposition rates [57, 58].

Available materials are wax-filled plastic adhesive material, proprietary nylon, and investment casting wax [54, 56]. The most popular modeling material is the ABS, which can be used on all current FDM machines. In addition, FDM technology can be used with polycarbonates, polycaprolactone (PCL), polyphenylsulfones, biodegradable plastic (PLA), and waxes (see Table 6.2).

Another material that has been developed to suit industrial standards is the ULTEM 9085 material. This has particularly favorable flame, smoke, and toxicity (FST) ratings that make it suitable for use in aircraft, marine, and ground vehicles. If applications require improved heat deflection, then the option would be to use the polyphenylsulfone (PPSF) material that has a heat deflection temperature at 264 psi of 189 °C.

It has to be noted that FDM works better with amorphous polymers rather than highly crystalline polymers, which are more suitable for PBF processes; this is because it works better with polymers in a viscous paste rather than in a lower viscosity form. All the materials used are nontoxic and are available in different colors.

Furthermore to the output of objects in plastic, material extrusion printers also can output other semi-liquid materials as in materials such as cheese or chocolate, and still in development are concrete printers that may in the future allow entire buildings (or large parts thereof) to be printed.

TABLE 6.2 FDM Modeling Materials

Material	Description
ABS, ABS plus	Best resolution for FDM and is environmentally stable—no appreciable warping, shrinkage, or moisture absorption. It offers a good blend of mechanical and esthetic properties.
PC	Is the most widely used industrial thermoplastic. Durable, it has high mechanical properties and is heat resistant. This material, while weaker than the normal PC, it is certified for use in food and drug packaging and medical device manufacture.
ABS/PC blend	Durable thermoplastic with a high gloss white or black color. It has superior mechanical properties and heat resistance of PC, excellent feature definition, surface appeal of ABS and highest impact strength.
PPSF/PPSU (polyphenylsulfone)	Highest heat and chemical resistance of all Fortus materials. It is a mechanically superior material with great strength, ideal for applications in caustic and high heat environments.
ULTEM	Extremely durable, high heat and chemical resistance; highest tensile and flexural strength; ideal for commercial transportation applications in airplanes, buses, trains, boats, etc.; high gloss gold/tan in color.

Another possible application of FDM is to develop ceramic part fabrication processes. In particular, FDM can be used to extrude ceramic pastes that can quickly solidify. The resulting parts can be fired using a high temperature furnace to fuse and densify the ceramic particles.

6.3.2.4 *FDM Machines* The FDM machines are generally lightweight assemblies with build volume capacities ranging roughly from 288 to 31K in.^{3} (4719–508K cm^3). The function of this hardware assembly of the FDM machines is to heat and pump the modeling material through the tip and onto the modeling surface to produce precise parts. Material cartridges supply plastic filament to the extrusion head. In the heated chamber, the head moves in the x- and y-axes while liquefying and depositing the material. The z stage plate moves down to give the part the third dimension [56].

Typically stepper motors or servo motors are employed to move the extrusion head and adjust the flow, and the head can be moved in both horizontal and vertical directions. The motor drives a set of feed wheels to provide a sufficient force to push the filament through the liquefier and tip. The feed wheels are driven in a counter-rotating direction to provide the torque to feed the filament, which acts as a piston [53]. In addition, FDM machines can possess a second nozzle to extrude support material and generate structures beneath overhanging sections. The support material is similar to the model material, but it is more brittle so that it may be easily removed after the model is completed. The FDM machine can build supports for any structure that has an overhang angle of less than 45° from horizontal.

Hence, while many of them use the phrase "FDM" to refer to this kind of 3D printing, actually only Stratasys makes FDM machines, which offers a very wide range from low-cost, small-scale, minimal variable machines through to larger, more versatile, and more sophisticated machines that are inevitably more expensive. Stratasys manufactures several lines of machines, including the product lines Dimension, uPrint, and Fortus as well as Hewlett Packard's Designjet line [56]. The Dimension machines under the Stratasys brand focus on the low-cost machines currently starting around US $15,000. Each Dimension machine can only process a limited range of materials, with only a few user-controllable parameter options. The uPrint machine is currently the smallest and lowest costing, with a maximum part size of 6 in × 8 in × 8 in. It has only one layer thickness setting and only one build material, with a soluble support system. There are two further machines that are slightly more expensive than the uPrint going upwards in size to 10 in × 10 in × 12 in with different layer thickness settings (0.25 and 0.33 mm) and ABS materials.

While Dimension FDM machines can be used for making parts for a wide variety of applications, most parts are likely to be used as concept models by companies investigating the early stages of product development. More demanding applications, such as for models for final product approval, functional testing models, and models for direct digital manufacturing (DDM), would perhaps require machines that are more versatile. Higher specification FDM machines are more expensive, not just because of the incorporated technology, but also because of the sales support, maintenance, and reliability. Stratasys has separated this higher-end technology through the subsidiary named FORTUS, with top-of-the-range models costing around $400,000. Further up the range are machines with increased size, accuracy, range of materials, and range of build speeds. The largest and most sophisticated machine is the FORTUS 900mc, which has the highest accuracy of all Stratasys FDM machines with a layer thickness of 0.076 mm. The build envelope is an impressive 36 in × 24 in × 36 in, and there are at least seven different material options.

On the other hand, since 2006 open source machines based on extrusion were released including Fab@Home, RepRap, and MakerBot. The Fab@Home, the first open source multi-material printer, is a syringe-based deposition system. An x-y-z gantry system moves a syringe pump across a 20 cm × 20 cm × 20 cm build volume at a maximum speed of 10 mm/s and resolution of 25 μm. Multiple syringes can be controlled independently to deposit material through syringe tips. That versatility allowed going beyond printing just in thermoplastics, as did the RepRap and most consumer-scale printers that followed. The range of materials that could be printed with a Fab@Home included hard materials such as epoxy, elastomers such as silicone, biological materials such as cell-seeded hydrogels, food materials such as chocolate, cookie dough, and cheese, engineering materials such as stainless steel, and active materials such as conductive wires and magnets [59–61]. From these solutions, MakerBot and RepRap use the best (cheapest and most flexible) control system, while Fab@Home has superior user friendliness (in hardware and software) as well as a hot-swappable tool head that accepts syringe, cutters, plastic tools, etc. The printer's multiple syringe-based deposition method allow for some of the first multi-material prints including direct fabrication of active batteries, actuators, and sensors, as well as esoteric materials for bioprinting and food printing.

6.3.2.5 Software/Firmware for FDM　The software/firmware controls the motion of the head assembly on the carriage and also the motion of the feed wheels. The major tasks of feed wheel control can be broken down into two major categories: steady-state and transient behavior. Steady-state pumping requires very accurate carriage and feed wheel control to assure precise geometry and road width. At the start or stop of the road, material flow is inherently different and requires different motor control to accommodate viscoelastic material behavior to precisely begin or end a road.

Before the carriage moves to start a new road, the feed wheels meter a small amount of material in anticipation of the carriage accelerating to the steady state. When the carriage moves, the flow is slowly turned on based on constant acceleration to the full pumping rate. Similarly, near the end of the road, the pumping rate decelerates prior to the actual end point, which creates a "starving" condition at the end point. The deceleration, acceleration, and prestart metering control values are dependent on the material viscoelastic properties at the application temperature. Each material has its own characteristics which must be programmed into the software/firmware [61].

6.3.2.6 Building Parameters in FDM　In order to be used as a part for serial production, the components must possess the required mechanical properties. The appropriate material characteristics including density, tensile modulus, flexural modulus, tensile strength, flexural strength, impact strength, and hardness will define the selection of the correct control process parameters for optimal model output and efficiency. The integration of material, hardware, and software in the FDM technology begins with the understanding of the basic requirements of the machine and ends with an operating procedure to choose these variables [57].

Several researchers, such as Said, Lee, Sood, Kumar, and Patel [62–66], have studied the effect of the FDM process control parameters that affect the properties of the FDM parts, and they discovered that the quality of FDM-made parts highly depends on three important process parameters such as bead width (layer thickness), raster orientation or angle, and air gap [67].

- *Bead (or Road) Width.* This is the thickness of the bead (or road) that the FDM nozzle deposits. It can vary from 0.012 in to 0.0396 in for Stratasys FDM machines.
- *Raster Orientation.* This is the direction of the beads of material (roads) relative to the loading of the part. In practice, most FDM parts are made with a criss-cross raster in which the orientation of the beads alternates from +45° to −45° from layer to layer. Other strategies to fill one layer are to generate all

contours or only contours to a specified depth. Coupled with the raster orientation, the width of the filament fill defines the toolpath.

- *Air Gap.* This is the space between the beads of FDM material. The default is zero, meaning that the beads just touch. It can be modified to leave a positive gap, which means that the beads of material do not touch. This results in a loosely packed structure that builds rapidly. It can also be modified to leave a negative gap, meaning that two beads partially occupy the same space. This results in a dense structure, which requires a longer build time.

Other variables involved in the process include part geometry, deposition geometry, deposition speed, material, flow control parameters, etc. Other parameters can be the temperature of the heating element for the model material, as it can control how molten the material is as it is extruded from the nozzle and the nozzle diameter, which is the width of the hole through which the material extrudes, defining the amount of material deposition.

6.3.2.7 Multiphase Jet Solidification (MJS)
Rapid prototyping (RP) systems were designed to reduce the time taken to develop new products, and, today, available RP systems work with different techniques using paper, polymers, and waxes. In order to fulfill the demand for the direct production of metallic prototypes for functional application and testing, the Fraunhofer Institute for Applied Materials Research (IFAM) has developed a new process, named "multiphase jet solidification (MJS)," which is able to produce metallic or ceramic parts. The MJS process uses low-melting alloys or a powder-binder mixture, which is squeezed out through a computer-controlled nozzle. Parts are manufactured layer by layer, and the "green parts" are debinding and sintered to reach final density.

In this process, metal or ceramic slurries are extruded using the metal injection molding (MIM) technique. The slurry is a mixture of wax and metal (low-melting alloys) or ceramic powder mixed in a ratio of about 50/50. It is kept in a heated container and a computer-controlled screw-activated plunger pumps it through an attached nozzle. Parts are manufactured layer by layer and the "green parts" are debinded and sintered to reach final density. The use of ceramic materials for the MJS process is under research at Rutgers University, New Jersey, USA [68].

6.3.2.8 Working Principle of MJS
The working principle of the MJS process is shown in Figure 6.7. The basic idea is comparable to the FDM process with regard to the layerwise deposition of molten material with low viscosity by a nozzle system. The main differences between both processes are in the raw material and the feeding system. For the MJS process, the material is supplied in different phases, for example, as a powder-binder mixture or as a liquefied substance instead of using wire material. In any case, the material is heated beyond its solidification point, squeezed out through a nozzle by a pumping system, and deposited layer by layer. The melted material solidifies when it comes into contact with the platform or the previous layer because of temperature, pressure decrease, and heat transference to the part and the environment. The contact of the liquefied material leads to partial remelting of the

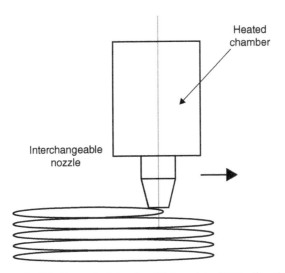

Figure 6.7 Working principle of multiphase jet solidification (MJS).

previous layer and to a good bonding of the layers. In order to make a functional metal part, the green part made via the deposition (typically 50–70% volume solid material) must then undergo a debinding step and a sintering step. After the sintering step, the part undergoes 10–16% linear shrinkage and the resultant part density is 95–98% [68–70].

In the FDM process, the wire material is transported into the heated chamber by two conveyor rollers, which make the discharge of the molten material possible. The MJS process uses a special feeding system suited to conveying powder material. The extrusion jet is mounted on a xyz table, which is controlled by a computer system. After one cross-section is produced in the x-y plane, the extrusion head moves in the z direction (slice heights: 0.5–0.1 mm), and the next layer is manufactured. The part is created layer by layer up to its final size. The part is produced from a 3D CAD model. Process parameters such as machining speed and material flow are added, and the control file for the machine is generated. The file is then downloaded to the controller of the machine and the build process is started.

The production of metallic and ceramic parts is the main aim of this development. Due to technological limitations, pure materials with a high melting point cannot be liquefied with the apparatus. If the melting point of the material is too high, additional powder-binder mixtures have to be used. The MJS process benefits from being able to process a wide variety of materials due to its similarities with MIM, the only limitations being that the materials must have a suitable viscosity (10–200 Pa's) and a binder melting temperature of less than 200 °C. The similarity to MIM also enables the MJS process to create parts of comparable material properties. Unfortunately, the use of an extrusion nozzle by the process limits its minimum feature size, layer thickness, and planar accuracy (±0.2 mm); as such, MJS is unable to make extremely small features and is suitable for producing medium-sized parts [69].

6.3.2.9   *Elements of the MJS Process*   The MJS machine was first commercialized as the RP-Jet 200 in 1996. Costs associated with this process are relatively low since a high powered energy source is not required for its operation. The main components of an MJS system are an *xyz*-computer-controlled positioning system, a heated chamber, and a jet system. The chamber is temperature-stabilized and is generally varied within a range of 70–220 °C. The material flow is controlled by a piston pressing the viscous material through the jet. The material is supplied as powder, pellets or bars [71].

6.3.2.10   *Low Temperature Deposition Manufacturing (LDM)*   LDM is another modified version of FDM developed by Xiong et al. [72] to overcome the heating and liquefying processing of materials. The system comprises a multi-nozzle extrusion process and a thermally induced phase separation process. This manufacturing method of solid freeform fabrication was proposed to fabricate poly(L-lactic acid)/(tricalcium phosphate) composite scaffolds for bone tissue engineering as it preserves the biological activity by fabricating custom pore-sized structures. LDM was recently used by Li et al. [73] to fabricate individualized tissue engineering PLGA/tricalcium phosphate (TCP) composite scaffolds based on alveolar bone defects. Mäkitie et al. [74] also assessed the viability of (PLGA/TCP) composite scaffold generated with LDM in 3D cell cultivation, *in vitro*.

6.3.2.11   *Bioextrusion*   Bioextrusion is the process of creating biocompatible and/or biodegradable components that are used to generate frameworks, commonly referred to as "scaffolds," which play host to animal cells for the formation of tissue. Such scaffolds should be porous, with micropores that allow cell adhesion and macropores that provide space for cells to grow. In the bioextrusion technique, a microneedle is employed as the extrusion nozzle where liquids, pastes, melts, solutions, hot melts, reactive oligomers, or dispersions, which are initially stored in a heated cartridge, are extruded into a temperature-controlled liquid dispensing medium. The dispensing medium induces solidification of the deposited material by cooling, heating, or through chemical reaction [75].

One common method of creating scaffolds is to use hydrogels. These are polymers that are water insoluble but can be dispersed in water. Hydrogels can, therefore, be extruded in a jelly-like form. Following extrusion, the water can be removed and a solid, porous media remains. Such media can be very biocompatible and conducive to cell growth with low toxicity levels. Overall, the use of hydrogels results in weak scaffolds that may be useful for soft tissue growth. When stronger scaffolds are required, melt extrusion seems to be the process of choice, as used to generate bone tissue.

There are a few commercial bioextrusion systems, such as Osteopore, which are used to create scaffolds to assist in primarily head trauma recovery. This machine uses a conventional FDM-like process with settings for a proprietary material, based on the biocompatible polymer, PCL.

Another developed machine is the 3D-Bioplotter by EnvisionTEC, which is an extrusion-based screw feeding technology designed specifically for biopolymers in bioextrusion. Biocompatible polymers suitable for tissue engineering are synthesized in relatively small quantities and are, therefore, only provided at a high cost. Lower temperature polymers can be extruded using a compressed gas feed, instead of a screw extruder, which results in a much simpler mechanism. Much of the system uses nonreactive stainless steel and the machine itself has a small build envelope and software specifically aimed at scaffold fabrication. The system uses one extrusion head at a time, with a carousel feeder so that extruders can be exchanged. This is particularly useful since most tissue engineering builds scaffolds with different regions made from different materials. Build parameters can be set for the control over the chamber temperature, feed rate, and plotting speed.

6.3.2.12 Other Extrusion-Based Processes Further novel and modified deposition techniques have been also introduced in the last three years. These methods were developed to increase manufacturing flexibility by enhancing deposition capability in achieving optimum scaffolds. Robocasting is an extrusion-based process in which a colloidal suspension, or ink, is extruded through a micron-sized nozzle in a defined trajectory to form a three-dimensional structure and is referred to in the literature as robotic deposition and direct-write assembly [75–77]. Recently, this technique has been used to fabricate porous β-TCP scaffolds with a controlled architecture [78].

Furthermore, in order to overcome filament preparation problem in FDM, a variation of FDM called precision extruding deposition (PED) for fabrication of bone tissue scaffolds were developed by Wang et al. [79]. In PED, material in pellet or granule form is fed into a chamber where it is liquefied. Pressure from a rotating screw forces the material down a chamber and out through a nozzle tip. This process is used to directly fabricate polycaprolactone (PCL) and (PCL–hydroxyapatite, HA) composite tissue scaffolds [75].

Other new methods can include the multihead deposition system (MHDS), screw extrusion system (SES), combined FDM and electrospinning (ESP) system, combined plotting and ESP, 3DF, and ESP, combined rapid freezing and plotting system, porogen-based extrusion system, and modified plotting system [75].

6.3.3 Medical Applications of Extrusion-Based Systems

6.3.3.1 Tissue Regeneration by Scaffolds Tissue engineering is an interdisciplinary field that typically involves the combination of cells and biomaterials to form tissues with the goal of replacing or restoring physiological functions, lost in diseased organs, through scaffolds. Tissue engineering scaffolds that mimic the complex architecture of native tissues have been more difficult to produce than conventional porous polymer scaffolds that support undirected cell adhesion and spreading within homogeneous and relatively large (millimeter scale) constructs [80].

The scaffold features typically, which are interconnected architectures, along with the material composition and porosity through design and fabrication, could be a critical factor in the future clinical success of tissue engineering. As an optimum scaffold has not been obtained yet, there are many research efforts to fulfill desired scaffold requirements by enhancing scaffolds design, material, and manufacturing processes. Actually, there are various conventional and manual-based techniques used for scaffold fabrication such as solvent-casting/particulate leaching, gas foaming, fiber bonding, phase separation, and emulsion freeze drying, which allow for limited control of pore size and shape but lack the sensitivity to precisely produce scaffolds with controlled internal architectural features [81]. In contrast, CAD-based rapid prototyping methods provide excellent spatial control over polymer architecture and have recently been applied to the fabrication of 3D tissue engineering scaffolds.

FDM is capable of acquiring complex internal organization by altering the direction of material deposition with each layer. Zein and Hutmacher used this method to produce biodegradable poly(ε-caprolactone) (PCL) scaffolds with various honeycomb geometries with finely tuned pore and channel dimensions of 250–700 μm. FDM can produce scaffolds composed of other biocompatible polymers and composites with demonstrated utility for various tissue engineering applications, such as for design and fabrication of custom-made Ti6Al4V scaffold architectures for orthopedic implants, for anatomical femoral and tibial cartilage, and furthermore, it is able to make replacement limbs and joints used in the medical field, as in tissue engineering.

While FDM exhibits high pattern resolution in the x-y plane, high processing temperatures limit the biomaterials that are compatible with the method. However, FDM capabilities are expanding with new developments such as MJS, a technique discussed in Section 6.3.2, which allows simultaneous extrusion of multiple melted materials.

6.3.3.2 Surgery Planning and Anatomical Models

One particular application is in the production of anatomical models, which can assist in the diagnosis and planning of a surgery. Experience from several branches of surgery suggests that the resulting benefits include a reduction in wound size, the elimination of additional surgery, an increased capability of performing a more complicated surgery, a raise in the potential of creating customized implants, and a surgery time reduction in between 17% and 60%. Anatomical models benefit not only the health service but the patient as well; studies have shown that patients recover more quickly, have less pain and anxiety, and are more satisfied with the outcome of their surgery.

Most medical models to date have used stereolithography (SL). However, FDM offers a better controlled process, which enables the model to be created in a single processing step. SL models require additional cleaning and curing under ultraviolet light, which increases the time to produce a model. Furthermore, the resin is toxic and expensive. Therefore, FDM is actually more suitable for a hospital environment than SL. The materials available for use in an FDM machine include a medical grade of acrylonitrile butadiene styrene (ABS) that can be sterilized using gamma radiation, and an investment cast wax [82].

6.3.3.3 Mandibular Reconstruction

Fabrication procedure of medical models for mandibular reconstructive surgery application is aided by the use of a computer-aided design (CAD) and FDM technique. Case studies of patients with mandibular defects are examined using CAD model construction including data acquisition from computerized tomography scan and data processing. The models are used in assisting the surgeons in the reconstruction planning. A significant improvement is found in the further application of FDM technology for the fabrication of the latter complex mandibular models used for reconstructive surgery [83].

6.3.3.4 Standing Frames and Mobility Aids
The application of FDM in standing frames and mobility aids for people with disabilities, such as spinal cord injuries, muscular dystrophy, multiple sclerosis, and cerebral palsy, aids not only in the fabrication, but also the past lengthy turnaround time for the models and the costs. A model that cost $1000 and took three weeks to make, now with FDM can be made for around $100 [84].

6.3.3.5 MRI Scanners
To produce the plastic parts used in MRI scanners, manufacturers have used conventional manufacturing techniques such as CNC machining, reaction injection molding (RIV), and room temperature vulcanization (RTV) molding. To accelerate design and manufacturing, reduce production costs, and make production more efficient for the parts' low production volumes, the shareholders are increasingly turning to DDM. One of the technologies is the FDM, which is particularly suitable for developing MRI components, as FDM can use production-grade thermoplastics, which withstand high heat, caustic chemicals, sterilization, and high-impact applications. The use of resins such as polycarbonate (PC) and PPSF can provide time, cost, and efficiency advantages in the fabrication of MRI prototypes [55].

6.3.3.6 Other Applications
In general, FDM product applications are in injection molding; fit, form, and function applications; and investment castings. In the first, the prototype injection molds are capable of being produced by FDM, as the stability of the material and the accuracy of the hardware allow models to be used as masters for it. FDM is also available for producing orthopedic, dental, and cranio-maxillofacial prototypes to check fit, form, and function at less cost and time than conventional processes. It is highly accurate also for surgical guides and prototypes that mimic the look and feel of the finished medical device, including clear and rubber-like materials, as it can output a range of textures from rigid to flexible. Finally, it can work for strong tooling, custom fixtures and end-use parts, as it works with production-grade materials, including high-performance thermoplastics.

The feasibility of FDM to fabricate porous customized freeform structures of medical-grade polymethylmethacrylate (PMMA) was investigated by Espalin et al. [85]. It was found that, by enabling the use of PMMA in FDM, medical implants such as custom craniofacial implants can be directly fabricated from medical imaging data improving the current state of PMMA use in medicine. Yen et al. [86] also employed FDM in the production of poly(D,L-lactide-*co*-glycolide) (PLGA) scaffolds filled with type II collagen and evaluated the cellular proliferation and matrix deposition of these hybrid scaffolds.

REFERENCES

[1] ASTM. *Standard Terminology for Additive Manufacturing Technologies*. ASTM; 2010. p F2792.

[2] Valkenaers H, Vogeler F, Ferraris E, Voet A, Kruth JP. 2013. A novel approach to additive manufacturing: screw extrusion 3D-printing. Proceedings of the 10th International Conference on Multi Material Micro Manufacture 359, 235–238.

[3] Vogeler F, Verheecke W, Voet A, Valkenaers H. An initial study of Aerosol Jet® printed interconnections on extrusion-based 3d-printed substrates. J Mech Eng 2013;59(11):689–696.

[4] Rezende RA, Silva JVL. 2012. Additive manufacturing and its future impact in logistics. 6th IFAC Conference on Management and Control of Production and Logistics, Vol. 6 Part 1, 277–282.

[5] Wohlers T. *Wohlers Report: Additive Manufacturing and 3D Printing State of Industry*. Wohlers Associates; 2012.

[6] Shulman H, Spradling D, Hoag C. *Introduction to Additive Manufacturing*. Ceramic Industry; 2012. p 15–19.

[7] Barnatt C. 2013. 3D Printing the Next Industrial Revolution, ExplainingTheFuture.com, ISBN-13: 978-1484181768, 26–40.

[8] AMCRC. *Additive Manufacturing Categories Processes and Materials*. Hawthorne, Australia: Advanced Manufacturing CRC Ltd; 2012. p 1–45.

[9] Nanyang Technological University, Singapore. 2001. Chapter 15. Ink Jet Printing in "Computer Peripherals", Lintech, Available in: http://www.lintech.org/comp-per/15INK.pdf

[10] Hutchings IM, Martin GD. *Inkjet Technology for Digital Fabrication*. West Sussex, UK: John Wiley & Sons; 2013. p 1–152.

[11] Somiya S. *Handbook of Advanced Ceramics: Materials, Applications, Processing, and Properties*. Oxford, UK: Academic Press, Elsevier; 2013. p 505–506.

[12] Green S. *The True Versatility of Inkjet 3D Printing in Action*. Stratasys Ltd.; 2012.

[13] Chang KH. *Product Manufacturing and Cost Estimating using CAD/CAE: The Computer Aided Engineering Design Series*. Elsevier, Oxford, UK: Academic Press; 2013. p 5.

[14] Pereira RF, Bartolo PJ. *Recent Advances in Additive Biomanufacturing*. Oxford, U.K.: Elsevier; 2014. p 12–13.

[15] Gibson I, Rosen DW, Stucker B. *Additive Manufacturing Technologies*. NY, USA: Springer; 2009

[16] THRE3D. 2014. How Binder Jetting Works. Available in: https://thre3d.com/how-it-works/binder-jetting.

[17] THRE3D. 2014. How Material Jetting Works. Available in: https://thre3d.com/how-it-works/material-jetting

[18] Hoag C, Holly Shulman H, Spradling D. *Introduction to Additive Manufacturing*. Ceramic Industry; 2012.

[19] Virginia Tech. 2014. Design, Research and Education for Additive Manufacturing Systems, Material Jetting. Available in: http://www.me.vt.edu/dreams/material-jetting/

[20] Chen ST. In: Kamanina N, editor, ISBN: 978-953-307-899-1. Inkjet Printing of Microcomponents: Theory, Design, Characteristics and Applications, Features of Liquid Crystal Display Materials and Processes. InTech; 2011. p 43–60.

[21] Uhland S, Holman RK, Morissette S, Cima MJ, Sachs EM. Strength of green ceramics with low binder content. J Am Ceram Soc 2001;84(12):2809–2818.

[22] Thomas GP. 2013. Materials Used In 3D Printing and Additive Manufacturing, AzOM, Available in: http://www.azom.com/article.aspx?ArticleID=8132#6

[23] Rozli MF. 2009. Design, Rapid Prototyping and Testing of the Wells Turbine, Thesis, Universiti Teknologi PETRONAS, 22.

[24] SHINING 3D TECH. 2009. Multi-Jet Modeling (MJM), Inkjet Printer 3D Printing, Rapid Prototyping & 3D Printing. Available in: http://www.shining3dscanner.com/en-us/MJM .html

[25] EFunda. 2014. Rapid Prototyping, Ink Jet Printing Techniques. Available in: https://www.efunda.com/processes/rapid_prototyping/inkjet.cfm

[26] Solid Concepts. 2014. PolyJet. Available in: http://www.solidconcepts.com/technologies/polyjet/

[27] Stratasys Ltd. 2014. Stratasys Redefines Product Design and Manufacturing with World's First Color Multi-material 3D Printer, Acquire Media. Available in: http://investors.stratasys.com/releasedetail.cfm?ReleaseID=821134

[28] LuXeXcel. 2010. Printoptical Technology, Available in: http://www.luxexcel.com/technology/

[29] Organovo. 2013a. The Bioprinting Process. Available in: http://www.organovo.com/science-technology/bioprinting-process

[30] Williams R. 2014. The next step: 3D printing the human body, The Telegraph. Available in: http://www.telegraph.co.uk/technology/news/10629531/The-next-step-3D-printing-the-human-body.html

[31] Organovo. 2013b. 3D Human Tissues. Available in: http://www.organovo.com/3d-human-tissues

[32] Silva JVL, Rezende RA. 2013. Additive Manufacturing and its future impact in logistics. 6th IFAC Conference on Management and Control of Production and Logistics, Volume # 6 Part# 1, 277–282.

[33] Park H, Park K. Biocompatibility issues of implantable drug delivery systems. Pharm Res 1996;13(12):1770–1776.

[34] Tarcha PJ, Verlee D, Hui HW, Setesak J, Antohe B, Radulescu D, Wallace D. The application of ink-jet technology for the coating and loading of drug-eluting stents. Ann Biomed Eng 2007;35(10):1791–1799.

[35] Cohen B. 2008. Drug-Eluting Stent Overview, Stent Center by Medtronic. Available in: http://www.ptca.org/articles/des.html

[36] Bártolo P et al. Biomedical production of implants by additive electrochemical and physical processes. CIRP Ann 2012;61:635–655. DOI: 10.1016/j.cirp.2012.05.005.

[37] Toh WS et al. Biomaterial-mediated delivery of microenvironmental cues for repair and regeneration of articular cartilage. Mol Pharmaceutics 2011;8:994–1001. DOI: 10.1021/mp100437a.

[38] Castro NJ, Hacking SA, Zhang LG. Recent progress in interfacial tissue engineering approaches for osteochondral defects. Ann Biomed Eng 2012;40(8):1628–1640. DOI: 10.1007/s10439-012-0605-5.

[39] Pereira RF, Almeida HA, Bártolo PJ. Biofabrication of hydrogel constructs. In: Coelho J, editor. Drug Delivery Systems: Advanced Technologies Potentially Applicable in Personalised Treatment, Advances in Predictive, Preventive and Personalised Medicine Volume 4. Berlin: Springer; 2013a. p 225–254.

[40] Sherwood JK et al. A three-dimensional osteochondral composite scaffold for articular cartilage repair. Biomaterials 2002;23:4739–4751. DOI: 10.1016/S0142-9612(02)
00223-5.

[41] Cui X et al. Direct human cartilage repair using three-dimensional bioprinting technology.
Tissue Eng Part A 2012;18(11–12):1304–1312.

[42] Coelho J. Drug Delivery Systems: Advanced Technologies Potentially Applicable in Personalised Treatment. Springer; 2013. p 243.

[43] Rezende RA, Pereira FDAS, Kasyanov V, Ovsianikov A, Torgensen J, Gruber P, Stampfl
J, Brakke K, Nogueira JA, Mironov V, Silva JVL. Design, physical prototyping and initial
characterisation of Lockyballs. Virtual Phys Prototyping 2012;7:287–301.

[44] Wake Health. 2014. Using Ink-Jet Technology to Print Organs and Tissue, Wake Forest
Institute for Regenerative Medicine. Available in: http://www.wakehealth.edu/Research/
WFIRM/Our-Story/Inside-the-Lab/Bioprinting.htm.

[45] Pereira RF, Barrias CC, Granja PL, Bartolo PJ. Advanced biofabrication strategies for
skin regeneration and repair. Nanomedicine 2013b;8(4):603–621.

[46] Langnau L. 2012. A closer look at extrusion-based 3D printers, Make parts fast, May 14,
2012.

[47] Reddy BV, Reddy NV, Ghosh A. Fused deposition modelling using direct extrusion.
Virtual Phys Prototyping 2007;2(1):51–60.

[48] Comb, J.W., Priedeman, W.R., Leavitt, P.J., Skubic R.L., Batchelder, J.S., 2005,
High-precision modeling filament, March 15 2005. US Patent 6,866,807.

[49] Masood SH, Song WQ. Thermal characteristics of a new metal/polymer material for
FDM rapid prototyping process. Assembly Autom 2005;25(4):309–315.

[50] Bellini A, Güçeri S, Bertoldi M. Liquefier dynamics in fused deposition. J Manuf Sci Eng
2004;126:237–246.

[51] Venkataraman N, Rangarajan S, Matthewson MJ, Harper B, Safari A, Danforth SC, Wu G,
Langrana N, Guceri S, Yardimci A. Feedstock material property—process relationships
in fused deposition of ceramics (FDC). Rapid Prototyping J 2000;6(4):244–252.

[52] Sun Q, Rizvi GM, Bellehumeur CT, Gu P. Effect of processing conditions on the bonding
quality of FDM polymer filaments. Rapid Prototyping J 2008;14(2):72–80.

[53] Chua CK, Leong KF, Lim CS. *Rapid Prototyping*. World Scientific; 2003. p 124. ISBN
9789812381170.

[54] Stratasys. 2013. FDM Technology. Available in http://www.stratasys.com/3d-printers/
technology/fdm-technology

[55] Winker R. *FDM Helps Build MRI Scanners*. Medical Design; 2009.

[56] Hiemenz, J., 2011, 3D Printing with FDM: How it Works, Stratasys in Control Design,
http://www.funtech.com/site/pdfs/SSYS-WP-3DP-HowItWorks-03-11.pdf (Accessed on
April 20, 2016), pp. 1–5.

[57] Malone E, Lipson H. 2006. Fab@Home: The Personal Desktop Fabricator Kit. Proceedings of the 17th Solid Freeform Fabrication Symposium, Austin TX, Aug 2006.

[58] Multiaxis. *Thermoplastic Materials in Fortus 3D Production Systems*. Cimtech, Inc.;
2013.

[59] Additive Manufacturing and 3D Printing State of the Industry. 2012. Annual Worldwide
Progress Report, ISBN 0-9754429-8-8

[60] Comb, JW, Priedeman, W.R., Tuley, P.W. 1994, *FDM Technology Process Improvements in Solid Freeform Fabrication Proceedings*, Marcus, H.L., University of Texas, Austin, USA1994, 42–50.

[61] Lipson H, Kurman M. *Fabricated: The New World of 3D Printing.* Wiley Press; 2013.

[62] Kumar Sood A, Phdar RK, Mahapatra SS. Improving dimensional accuracy of fused deposition modeling processed part using grey Taguchi method. Mater Des 2009;30:4243–4252.

[63] Lee BH, Abdulla J, Khan ZA. Optimization of rapid prototyping parameters for production of flexible ABS object. J Mater Process Technol 2005;169:54–61.

[64] Patel JP. 2012. An experimental investigation of process variable influencing the quality of FDM fabricated polycarbonate parts, M.Tech–Thesis.

[65] Said OE, Foyos J, Noorani R, Mandelson M, Marloth R, Pregger BA. Effect of layer orientation on mechanical properties of rapid prototyped samples. Mater Manuf Processes 2000;15(1):107–122.

[66] Sood AK, Ohdar RK, Mahapatra SS. Experimental investigation and empirical modeling of FDM process for compressive strength improvement. J Adv Res 2012;3(1):81–90.

[67] Brock JM, Montero M, Odell D, Roundy S. 2000. ME 222 Final Project, 2000, Fused Deposition Modeling (FDM) Material Properties Characterization.

[68] Greulich M, Greul M, Pintat T. Fast, functional prototypes via multiphase jet solidification. Rapid Prototyping J 1995;1(1):20–25.

[69] Greul M, Pintat T, Greulich M. *Rapid Prototyping of Functional Metallic and Ceramic Parts Using the Multiphase Jet Solidification (MJS) Process.* Washington, D.C.: Advances in Powder Metallurgy and Particulate Materials; 1996. p 7.281–7.287.

[70] Yagnik D. 2011. Rapid manufacturing—the next industrial revolution. International Conference on Current Trends in Technology, Vol. 2, 08–10 December, 2011.

[71] Oechsner A, Silva FM, Altenbach H. *Characterization and Development of Biosystems and Biomaterials.* Heidelberg, Germany: Springer-Verlag; 2013.

[72] Xiong Z, Yan Y, Wang S, Zhang R, Zhang C. Fabrication of porous scaffolds for bone tissue engineering via low-temperature deposition. Scr Mater 2002;46(11):771–776.

[73] Li J, Zhang L, Lv S, Li S, Wang N, Zhang Z. Fabrication of individual scaffolds based on a patient-specific alveolar bone defect model. J Biotechnol 2011;151(1):87–93.

[74] Mäkitie AA, Yan Y, Wang X, Xiong Z, Paloheimo KS, Tuomi J, Paloheimo M, Salo J, Renkonen R. In vitro evaluation of a 3D PLGA—TCP composite scaffold in an experimental bioreactor. J Bioact Compat Polym 2009;24:75.

[75] Abdelaal OA, Darwish SM. Fabrication of tissue engineering scaffolds using rapid prototyping techniques, world academy of science. Eng Technol 2011;59:11–27.

[76] Hoelzle DJ, Alleyne AG, Johnson AJW. Micro robotic deposition guidelines by a design of experiments approach to maximize fabrication reliability for the bone scaffold application. Acta Biomater 2008;4:897–912.

[77] Miranda P, Saiz E, Gryn K, Tomsia AP. Sintering and robocasting of β-tricalcium phosphate scaffolds for orthopaedic applications. Acta Biomater 2006;2:457–466.

[78] Martínez-Vázquez FJ, Perera FJ, Miranda FH, Pajares P, Guiberteau AF. Improving the compressive strength of bioceramic robocast scaffolds by polymer infiltration. Acta Biomater 2010;6:4361–4368.

[79] Wang F, Shor L, Darling A, Khalil S, Sun W, Güçeri S, Lau A. Precision extruding deposition and characterization of cellular polycaprolactone tissue scaffolds. Rapid Prototyping J 2004;10:42–49.

[80] Agrawal CM, Ray RB. Biodegradable polymeric scaffolds for musculoskeletal tissue engineering. J Biomed Mater Res 2001;55(2):141–150.

[81] Das S, Hollister S. In: Buschow KHJ, Cahn RW, Flemings MC, Ilschner B, Kramer EJ, Mahajan S, editors. *Tissue Engineering Scaffolds (invited), Encyclopedia of Materials-Science and Technology*. Elsevier; 2002.

[82] Meakin JR, Shepherd DET, Hukins DWL. Fused deposition models from CT scans. J Radiol 2004;77:504–507.

[83] Kouhi E, Masood S, Morsi Y. Design and fabrication of reconstructive mandibular models using fused deposition modeling. Assembly Autom 2008;28(3):246–254.

[84] Altimate Medical, Stratasys Case Studies. 2013. Available in: http://www.stratasys.com/resources/case-studies/medical/altimate-medical

[85] Espalin D, Arcaute K, Rodriguez D, Medina F, Posner M, Wicker R. Fused deposition modeling of patient-specific polymethylmethacrylate implants. Rapid Prototyping J 2010;16(3):164–173.

[86] Yen HJ, Tseng CS, Hsu SH, Tsai CL. Evaluation of chondrocyte growth in the highly porous scaffolds made by fused deposition manufacturing (FDM) filled with type II collagen. Biomed Microdevices 2009;11(3):615–624.

7

CERTIFICATION FOR MEDICAL DEVICES

CORRADO PAGANELLI, MARINO BINDI, LAURA LAFFRANCHI, DOMENICO DALESSANDRI, AND STEFANO SALGARELLO

Department of Medical and Surgical Specialties, Radiological Sciences, and Public Health, University of Brescia, Brescia, Lombardy, Italy

ANTONIO FIORENTINO

Department of Mechanical and Industrial Engineering, University of Brescia, Brescia, Lombardy, Italy

GIUSEPPE VATRI

Major Prodotti Dentari Spa, Moncalieri, TO, Italy

ARNE HENSTEN

Faculty of Health Sciences, UiT The Arctic University of Norway, Tromsø, Norway

7.1 INTRODUCTION

Medical devices and materials are subjected in most countries to legislations intended to keep under control the marketing of devices and materials, considering the quality, the performances for the intended use, and the safety for the patient and the user. Individual countries—such as United States, or Japan—or politically or economically grouped countries—such as the European Union and MERCOSUR—have developed similar legislations. The similarities, however, are not always more important than the differences [1–16].

Biomedical Devices: Design, Prototyping, and Manufacturing, First Edition.
Edited by Tuğrul Özel, Paolo Jorge Bártolo, Elisabetta Ceretti, Joaquim De Ciurana Gay, Ciro Angel Rodriguez, and Jorge Vicente Lopes Da Silva.

A world instrument for the harmonization and standardization of at least some aspects of the medical devices regulatory processes was imagined in 1993, giving birth to the *Global Harmonization Task Force* (GHTF) "an international partnership between medical device regulatory authorities and the normalized industry, one that would aim to achieve harmonisation in medical device regulatory practices." "The purpose of the GHTF was to encourage convergence in regulatory practices related to ensuring the safety, effectiveness/performance and quality of medical devices, promoting technological innovation and facilitating international trade, and the primary way in which this was accomplished was via the publication and dissemination of harmonized guidance documents on basic regulatory practices." (www.imdrf.org/ghtf).

The GHTF's two decennial works have left a list of primary reference documents, largely used by regulatory authorities. From 2011, the GHTF ceased to exist and was substituted by the *International Medical Device Regulatory Forum* (IMDRF). The members represent the main world areas interested in premarketing registering, or approving, or certifying medical devices. Members of the management committee, at a national or supranational level, are

- the United States Food and Drug Administration (US FDA),
- the European Union Commission Directorate General Health and Consumers (EU),
- Pharmaceuticals and Medical Devices Agency (PMDA)—Japan,
- Health Canada—Canada,
- National Health Surveillance Agency (ANVISA)—Brazil,
- Therapeutic Goods Administration (TGA)—Australia.

The IMDRF maintains and develops a list of documents, which are taken as suggestions or the basis for standard and legislation development. The heavy influence of the EU mentality on market regulation and control is evident in many of those documents. Today, the approved and valid documents (not superseded, not substituted by an International Standard) are regrouped in five study areas:

1. *Premarket Evaluation.* Principles of classification, principles of safety and performance, principles of conformity assessment, use of the standards, etc.
2. *Postmarket Surveillance/Vigilance.* Principles and requirements of surveillance, adverse effects reporting, notifications and reports, filing and information exchange between regulatory agencies, etc.
3. *Quality Systems.* Processes validation, risk management activities, control of products and services obtained from suppliers, etc.
4. *Auditing.* Guidelines for auditing quality systems and for regulatory auditing of medical devices.
5. *Clinical Safety/Performance.* Clinical investigations, clinical evaluation, clinical evidence, performance evaluation, etc.

Other documents are under development.

7.2   THE MEDICAL DEVICES APPROVAL, REGISTRATION, OR CERTIFICATION

It is a complex process where, before (*premarket*) and after (*postmarket*) being marketed, a medical device or material has to be evaluated for performance and safety under the legislation provisions. Notwithstanding the local, also large, differences, the process always shows six cardinal waypoints.

1. The product is assigned to taxonomy of risk (for instance, Europe has four classes, from the lowest to the highest risk, 1, 2a, 2b, 3). The assessment, validation, and auditing procedures can be differentiated on the basis of the taxonomy (for instance, in Europe the Class 1 products are self-assessed by the manufacturer, while the other classes require a third-party validation).
2. A list of safety and performance principles or requirements is generated and maintained by the legislation or by a regulatory agency. The manufacturer submits a technical documentation (*Submission document*) for demonstrating the conformity of the product or service to the safety and performance requirements, either to a governmental agency (FDA), or to a third-party body (EU)—in some cases, a self-certification is admitted.
3. The receiving body(ies) audit(s) the manufacturer's quality management systems (good manufacturing procedures) and/or the technical documentation submitted and its implementation within the manufacturer's processes. At the end of the process, the receiving body releases an official registration or approval or certification for marketing the product or service.
4. The regulatory authority enters the manufacturer and their products or services into the national register of the marketed medical devices.
5. The manufacturer implements and keeps active a system of postmarket surveillance able to assure the continuous conformity of devices and the treatment of incident or dangerous nonconformities. The market actors (health professionals, local authorities, manufacturers) report to the regulatory authority the possible adverse effects. The regulatory authorities may exchange the information and keep a full list of known adverse effects.
6. The manufacturers are periodically audited for assuring the maintenance of the conformity to the safety and performance principles and requirements, through the quality management systems and the technical documentation updating and implementation.

7.3   THE PREMARKET KEY ACTIVITY: THE DEMONSTRATION OF THE CONFORMITY TO THE SAFETY AND PERFORMANCE REQUIREMENTS

It's understandable that, every individual country can have local specific legislation requirements and procedures so that a unified approval road for medical devices and materials cannot be traced. However, the convergence work within the IMDRF is producing a guideline for submitting an application for premarketing approval. This guideline defines an end document reporting what is necessary for approval or certification, taking into account universal sections and national specificities. The IMDRF proposal for a submission document, gives the following list of universal application headings, each one specified for the possible subheadings. *The Table of contents of a submission document for a regulated product can be considered, in addition to the bureaucratic and official contents, a complete checklist for documenting the premarket conformity to the safety and performance principles (Table 7.1).*

TABLE 7.1   Checklist for Documenting the Premarket Conformity to the Safety and Performance Principles

A. Nonclinical Section

Risk management	A paper reporting in full or in summary the risks identified during the risk analysis process and how they are controlled
Standards and conformity to nonclinical studies	Applied standards and declaration of conformity and/or (if available) third-party certifications
	Physical and mechanical characterization
	Chemical characterization
	Electrical safety and electromagnetic compatibility
	Radiation safety
	Software/firmware
	Biocompatibility and toxicological evaluation
	Immunological testing
	Pyrogenicity evaluation
	Biological safety
	Sterilization validation
	Animal testing
	Human factors/usability

Nonclinical bibliography
Safety and performance studies to support combination products
Expiration period and package validation
Other nonclinical evidence

B. Clinical Section

Overall clinical evidence	Device specific clinical trials
	Clinical literature review

Other clinical evidence

C. Information to User and Patient
Information on packages and instruction
Labeling for physicians
Labeling for patients
Technical manuals

D. Quality Management of Critical Areas
Quality system management document and procedures

Source: IMDRF WG(PD1)/N8R1.

7.4   THE POSTMARKET KEY ACTIVITY: THE SURVEILLANCE

The postmarket surveillance is the less harmonized area, almost every regulatory agency or legislation has individual and differently extended procedures. In general, the *surveillance* is intended as the pro-active work of manufacturers or regulatory agencies for gaining information about performance and safety of devices placed on the market. The term *vigilance* refers to the regulatory agencies passive activity of knowing, following, and recording incidents related to medical devices.

The manufacturer's activity is that of

- *Ordinary Surveillance.* Systematically gather the information coming from any road (market information, similar products, etc.), or searched (literature, customer complaints, etc.); analyze them to determine whether preventive or corrective actions are needed to maintain the compliance to the performance and safety requirements.
- *Extraordinary Surveillance.* Deeply search and analyze all cases where the performance or safety of devices could be considered impaired, leading to incidents, product recalls, or field safety notices (urgent communications sent out to reduce a serious risk associated with the device's use); keeping the competent authorities informed.

Postmarket surveillance includes also the clinical surveillance. Clinical data coming from the postmarket surveillance are critical to update the premarket clinical evaluation during the life cycle of a medical device. In addition, clinical follow-up plans can answer specific questions about performance and safety of in-use devices.

Postmarket surveillance can normally be ordered, inspected, and assessed by regulatory agencies.

7.5   THE ROLE OF THE QUALITY MANAGEMENT SYSTEMS

A quality management system is a system that leads and enables the company to keep the quality under control, or that is to say, with reference to the level at which a set of intrinsic distinctive elements (characteristics) satisfies a set of requirements. If these requirements are of a regulatory type, the quality management system can also keep under control the full compliance to the regulatory requirements. A quality management system can also easily be designed around the pivotal areas of the medical device's performance and safety (design and development, processes validation, nonconformities management, postmarket surveillance, etc.).

Because of this potentiality, a quality management system implementation (and/or a third-party certification) is often required by the regulatory agencies or by the legislations. The inspecting bodies use the quality management system to assess the continuous meeting of regulatory requirements.

The key standard for the medical devices is the ISO 13485:2003 *Medical devices—Quality management systems—Requirement for regulatory purposes.* It is accompanied by the guideline ISO/TR 14969:2004 *Medical devices—Quality management systems—Guidance on the application of ISO 13485:2003.*

7.6 THE VERIFICATION AND THE AUDITING

The regulatory authority, when requested, performs or allows to be performed by a third-party, the verification of the submission document. This verification has two main sections:

1. *Verification of Nonclinical and Clinical Evidences.* This activity requires special competencies and knowledge, not always fully available or simply referable to the legislation or to other normative provisions—for instance, the risk management of a new product, or the evaluation of the clinical evidence for new products. In these cases, the interaction between the professionals stating and evaluating the activity plans and the conclusions, and the experts within the regulatory body or the accredited third parties, is the preferred road.

2. *Full Auditing of the Manufacturer Activities of Quality Management, Design, Manufacturing, and Surveillance.* In the modern concepts of organization and management, auditing is a key activity. Auditing is defined as a "systematic independent and documented process for obtaining audit evidence and evaluating it objectively to determine the extent to which the audit criteria are fulfilled." (ISO 19011:2002). Auditing is then a systematic examination of data, records, operation, and performances of a body or an organization or an activity, with some stated purposes, as evaluating the compliance to a regulatory system, or the existence of stated plans and the compliance to them for processes, etc. Under the regulatory point of view, it is "The audit of a quality management system to demonstrate conformity with quality management system requirements for regulatory purposes" (GHTF/SG4/N39R21: 2010).

The verification has the main scope to verify the compliance of all activities to the regulatory requirements. So the auditing should be intended rather as a system improving activity than as a control from some authority.

A regulatory body auditing has to be regularly performed (*surveillance auditing*) at stated intervals, to confirm the ongoing conformity with the regulatory and quality system requirements. Surveillance audits are also requested as a fundamental part of the manufacturer's quality management system (*internal audits*).

Auditors of medical device activities are requested to have training and/or a professional history in medical devices, to be trained in all the relevant areas (as risk management, manufacturing processes, process validation, etc.), and to receive the accreditation after an adequate training in auditing.

Auditing guidelines are available as ISO standard (as ISO 19011:2011, *Guidelines for auditing management systems*) and as medical devices specialized documents at IMDRF (GTHF/SG4).

7.7 THE ROLE OF THE STANDARDS

Many steps of the approval/certification process are assisted by International Standards issued by the International Standard Organization (ISO). A standard is a document providing guidelines, specifications, and requirements that can be used for ensuring that activities, products, or services fit for their purpose. The interested groups (e.g., universities, manufacturers, traders, professionals) write down the ISO standards for the benefit of the public. The aim of the standards is to define minimum requirements and optimized procedures to state and evaluate the characteristics, the performances, and the procedures for products, services, materials, and processes. In the case of the medical standards, their aim in providing protection for the patient and the medical personnel is evident. The ISO states that "conformity to International Standards helps reassure consumers that product are safe, efficient and good for the environment." Due to the "voluntary consensus among interests of doers" philosophy informing them, it is possible that a standard, in some case, states requirements that reflect the basic common level. The process of periodical revision assures that level to be rethought regularly. ISO standards are periodically revised to adjust them to the newest technologies and to incorporate new matters and resources. Due to the long process of writing and updating, sometimes a standard cannot reflect the last state of the art. However, they have at large demonstrated their utility and there is a wide consensus (when not stated by legislations) in accepting them as a mean to demonstrate the compliance to the safety and performance requirements.

National Standardization Bodies can, as in Japan (JSA) or United States (AINSI/ADA), maintain national standards, which can more or less differ from the ISO ones. The historical trend indicates that differences and presences/absences will be progressively eliminated in favor of a world standardization. Within the European legislation, they can be used to specify the "Essential Requirements" conformity, requested to market a product or a service.

In the medical devices field, at least three different types of standards are available, which can be interpreted as different levels of standardization. As an example, *dental medical devices* are used to explicate the structure.

At the ground level, there are the *dental products or materials performance standards*. They give the requirements needed to evaluate the minimum performance for a given product. For instance, the Standard ISO 22112:2005, *Dentistry—Artificial teeth for dental prostheses*, states a set of technical and information requirements for synthetic polymer- and ceramic-based artificial teeth. Not all devices or materials covered by a product standard, of course, are considered medical devices in any jurisdiction, for instance, ISO 15912:2006, *Dentistry—Casting investments and refractory die materials*.

On the first floor, there are the *dental medical devices standards*. They look specifically at the generality of the dental devices, considered as a cluster of the medical devices. For instance, ISO 7405:2008, *Dentistry—Evaluation of biocompatibility of medical devices used in dentistry*, which applies and specifies to the dental cluster the requirements of ISO 10993 (see below) or ISO 1942:2009, *Dentistry—Vocabulary*.

On the second floor, there are *the medical device standards*. They are applied to the generality of the medical devices, considered as a special cluster of products and services, but requiring some added attention and more restrictive specification as health-protecting items. Between them, for instance, ISO/TS 19218-1-2:2011 *Medical devices—Hierarchical coding structure for adverse events*.

7.8 EXAMPLES OF APPROBATION/CERTIFICATION ROADS IN SOME WORLD AREAS

7.8.1 European Union

The classification of a medical device determines the conformity assessment procedures a manufacturer can choose to ensure that the device is adequately assessed: higher classification devices must undergo more stringent conformity assessment procedures than lower classification devices.

The EU has multiple directives to cover medical devices:

- Medical Device Directive (MDD) 93/42/EEC,
- Active Implantable Medical Device Directive (AIMDD) 90/385/EEC.

EU Directive 2007/47/EC, introduced on September 5, 2007 in the European Parliament, made significant amendments to the MDD and AIMDD. The changes introduced by the new directive are fully effective from March 21, 2010.

These directives categorize devices into four classes (I, IIa, IIb, III) on the basis of increasing risks associated with the intended use.

7.8.2 United States of America

The US Food and Drug Administration (US FDA) has established classifications for approximately 1700 different generic types of devices and grouped them into 16 medical specialties referred to as panels. Each of these generic types of devices is assigned to one of three regulatory classes (I-II-III) based on the level of control necessary to assure the safety and effectiveness of the device and it defines the regulatory requirements for a general device type.

7.8.3 Japan

Pharmaceuticals and medical devices (PMDs) are regulated by the PMDA under a system that grants marketing approval only when it is indicated that effectiveness and safety are guaranteed.

MDs are classified by design complexity, use characteristics, and risk assessment. The level of control, supervision, and data content needed to support the product depend on its classification.

7.8.4 Australia

The conformity assessment evidence needs to be registered with the TGA for all medical devices, except Class I nonmeasuring and nonsterile medical devices; however, an Australian Declaration of Conformity and supporting evidence in a suitable technical file must be maintained by the manufacturer for Class I medical devices and must be provided to the TGA if requested.

Conformity assessment evidence is also not required for some systems and procedure packs; however, the manufacturer must hold and maintain evidence that each medical device in the system or procedure pack meets the Essential Principles and that the relevant conformity assessment procedures have been applied.

The Australian regulatory framework, introduced in October 2002, has many similarities with that adopted by the European Union (EU); so, as the Australian and the EU regulatory requirements are similar, the TGA has determined that certificates issued by EU Notified Bodies may be accepted as conformity assessment evidence for the supply of devices in Australia. There are medical devices that are exceptions to this determination. At the same time, the US system does not align with the Australian regulatory framework.

7.8.5 Brazil

Medical devices are regulated by Law No. 6360 of 1976, decree 74.094/97. The Brazilian rules are identical to the European rules except for Rules 8 and 13. Resolution RDC-185 of October 22, 2001, is the main resolution for medical devices. This resolution describes the required documents for registering a product and it contains a registration protocol; resolution RDC No. 206 of November 2006 describes the requirements for registering *in vitro* diagnostic devices. Brazil has adopted the Universal Medical Device Nomenclature System (UMDNS).

7.8.6 Canada

The EU has four classes of medical devices, which generally correspond to Canada's four classes (I–IV).

The classification rules for devices other than IVDDs are close to, but not identical with, the European classification rules.

Any person who imports or sells a medical device in Canada, and any manufacturer of a Class I device who does not import or distribute solely through a person who holds an Establishment License, must hold an Establishment License (which is valid for one year). Retailers, health-care facilities, and manufacturers of Class II, III, and IV devices are exempt from this requirement. Nevertheless, prior to selling a device in Canada, manufacturers of Class II, III, and IV devices must obtain a Medical Device License; Class I devices are exempt from device licensing requirement.

7.9 IN-DEPTH STUDIES

7.9.1 Essentials of Safety and Performance Principles *(Source:* IMDRF Document GHTF/SG1/N68:2012)

7.9.1.1 General Principles "Medical devices should be designed and manufactured in such a way that, when used under the conditions and for the purposes intended and, where applicable, by virtue of the technical knowledge, experience, education, or training, and the medical and physical conditions of intended users, they will perform as intended by the manufacturer and not compromise the clinical condition or the safety of patients, or the safety and health of users or, where applicable, other persons, provided that any risks which may be associated with their use constitute acceptable risks when weighed against the benefits to the patient and are compatible with a high level of protection of health and safety."

The solutions adopted for designing and manufacturing should conform to safety principles. If a risk reduction is required, it has to be performed using the principles listed in the priority order:

- "identify known or foreseeable hazards and estimate the associated risks arising from the intended use and foreseeable misuse;
- eliminate risks as far as reasonably practicable through inherently safe design and manufacture;
- reduce as far as reasonably practicable the remaining risks by taking adequate protection measures, including alarms; and
- inform users of any residual risks."

Devices should achieve the performance intended by the manufacturer and be suitable for their intended purpose.

Characteristics and performances referred earlier should be maintained during the device lifetime and should not be affected to a degree dangerous for the health and safety of the patient and user.

Characteristics and performances referred earlier should be not adversely affected by transport and storage conditions, taking into account the instruction provided by the manufacturer.

"All known and foreseeable risks, and any undesirable effects, should be minimised and be acceptable when weighed against the benefits of the intended performance of medical devices during normal conditions of use."

7.9.1.2 Chemical, Physical, and Biological Properties The device should be manufactured and designed in such a way as

- to pose a special attention to the toxicity and flammability of the materials used;
- to pose a special attention the compatibility of the materials used and the biological tissues, cells, and body fluids;
- to pose a special attention to the performances of the materials used;

- to minimize the risk posed by contaminants and residues;
- to be used safely with substances, gases, and materials with which they come in contact;
- to be compatible with the medicinal products they are intended to administer;
- to reduce as reasonably possible the risks from leaking or leaching substances;
- to reduce as reasonably possible the risks from unintentional ingress or egress of substances in the environment in which is has to be used.

7.9.1.3 Infection and Microbial Contamination
The device should be manufactured and designed in such a way as

- to eliminate or reduce as reasonably possible the risk of infection;
- to provide easy handling;
- where necessary, to reduce any microbial leakage or any microbial exposure during use, and to prevent microbial contamination;
- to maintain the claimed microbiological state during the transport and storage;
- to assure the claimed sterile state through appropriated processes of manufacturing and packaging, and to maintain that state during the transport and storage;
- when a microbiological or sterile state is claimed, to be manufactured with validated methods;
- when it is intended to be sterilized, to be manufactured in controlled environmental conditions;
- when nonsterile, to be able to maintain cleanliness and integrity;
- when it is intended to be sterilized prior to use, to minimize the risk of microbial contamination;
- to help the differential identification of similar products in different sterile/nonsterile conditions.

7.9.1.4 Devices Incorporating Medicinal Substances or Materials of Biological Origin
(A full treatment of the matter is yet missing.)
Where a device incorporates as an integral part a medicinal substance or drug, the safety and performance of the whole has to be verified, as those of the medicinal material.

In some jurisdictions (for example, Australia), products incorporating tissues, cells, and substances of animal, human, or microbiological origin can be considered medical devices. In this case, selection, processing, preservation, testing, and handling of tissues, cells, or substances must assure optimal safety for patient and user. A special attention has to be paid to validated methods of elimination or inactivation of viruses or transmissible agents.

7.9.1.5 Environmental Properties
Devices intended for use in combination or in connection with other devices or equipment have to be evaluated also as a combination; the connections must be safe.

The device should be manufactured and designed in such a way as to reduce as reasonably possible the risk(s)

- of injury to patient or users;
- of use errors;
- associated with external influences or environmental conditions, magnetic fields, temperature, etc.;
- associated with contacts with materials and substances to which it is exposed or which can penetrate the device;
- of negative interactions between the device software and the use environment;
- of reciprocal interference with other devices;
- arising from aging, impossibility of maintenance, loss of accuracy of control systems;
- of fire and explosion.

The device should be manufactured and designed in such a way as

- adjustment and maintenance, where necessary, can be done safely;
- the disposal of any waste substance can be done safely.

7.9.1.6 *Diagnostic or Measuring Functions*   The diagnostic devices or those with a measuring function should be manufactured and designed in such a way as

- to provide sufficient accuracy, precision, and stability;
- to apply ergonomic principles to measurements, monitoring, and display scales;
- to express values in unites accepted and understood by the user.

7.9.1.7 *Protection Against Radiation*   The device should be manufactured and designed in such a way as

- to reduce as reasonably possible the exposure of patient, users, and others;
- to make possible for the user to control the emissions, and to give warnings of emissions, when the emission levels are hazardous;
- to reduce as reasonably possible the exposure to unintended radiations.

The devices emitting ionizing radiations should be manufactured and designed in such a way as

- to ensure that the radiation parameters can be varied taking into account the intended use;
- to give appropriate output quality for the diagnostic purposes while minimizing the exposure;
- to enable monitoring and control of delivered dose and to control the emission parameter, for therapeutic radiology.

7.9.1.8 Medical Device Software Device incorporating programmable systems or software, or stand-alone software, should ensure reliability, repeatability, and performance according to the intended use. Software in itself must be validated according the state of the art. Single fault condition events must find means to be eliminated or reduced.

7.9.1.9 Active Medical Devices In the event of single fault condition, consequent risks should be eliminated or reduced as reasonably possible.

Where the safety of the patient depends on power supply, the state of the power supply should be determinable and signaled.

Where devices are intended to monitor the patient parameters, appropriate alarm systems should advice of potentially dangerous situations.

The risk of active or passive influence of electromagnetic interferences should be adequately eliminated. Risks from accidental electrical shocks should be avoided.

7.9.1.10 Mechanical Risks The device should be manufactured and designed in such a way as

- to protect the patient and user against mechanical risks;
- to reduce the risk coming from vibrations generated by the device;
- to reduce at the lowest level the risk coming from the noise emitted by the device;
- to minimize the risks coming from handling of connections energy supplies and the risks of connection errors;
- to assure that potentially dangerous temperatures are not attainable on the accessible parts of the device.

7.9.1.11 Supplied Energy or Substances The device intended to supply the patient with energy or substances should assure the amounts set accurately enough to guarantee the patient and user safety. The device should prevent or indicate inadequacies of delivering and prevent the accidental release of dangerous level of energy or substances. Functions of controls and indicators should be clearly specified on the device and understandable to the user.

7.9.1.12 Medical Devices Intended to be Used by Nonprofessionals They should perform appropriately taking into account skills and means available to nonprofessionals. They should reduce the risk of errors in handling, operating, and interpreting. Where needed and possible, they should provide procedures for verifying that the product performs appropriately.

7.9.1.13 Labels and Instructions Users should be provided with the information needed to identify the manufacturer, the device, the intended use, and the instruction to use it safely and to obtain the intended performance, taking into account the user knowledge and training.

7.9.1.14 Clinical Evaluation The demonstration of conformity with these essential principles must include a clinical evaluation. The clinical evaluation should review clinical investigation reports, literature reports or reviews, and clinical experience. The clinical evaluation should establish that a favorable risk/benefit ratio for the device exists.

7.9.2 Essentials of the Risk Management (*Source*: IMDRF Document GHTF/SG3/N15R8)

7.9.2.1 Key Terms

- *Harm*. Damage to the health, to property, or to environment.
- *Hazard*. Potential source of harm. The measure of its possible consequences is called *severity*.
- *Risk*. Combination of the probability of occurrence of a harm and its severity. The process of assigning a value to the combination is called risk *estimation*. The risk remaining after taking measures to reduce the risk is called *residual risk*.

7.9.2.2 Key Concepts

- The *criteria for the risk acceptability* (normally, grounded in some indexes) that will be used in the risk management process.
- The idea that the *full life-cycle of the device* has to be examined within the risk management process.

7.9.2.3 International Standards A detailed guide to the *Risk Management* process is given by the Standard *ISO 14971:2007 Medical devices—Application of risk managements to medical devices.*

The standard designs the risk management as a flow procedure executing a stated planning:

I. *Risk Analysis*. It is the process by which the intended uses, the users, the environment, the possible misuses, and the safety characteristics of a device are identified and analyzed. Then, all the possible related hazards are identified and the connected risks estimated.

II. *Risk Evaluation*. The following step, for every risk, is the evaluation if the risk as estimated could be accepted or has to be reduced. The risks to be reduced are submitted to an activity of risk control.

III. Risk Control.

Measures. For reducing the risk, the manufacturer has three basic options: to make the device inherently safe with a new design, to adopt protective measures, and to provide added information for safety.

Residual Risks Evaluation. After the implementation of the risk control measures, a new risk evaluation is performed, taking into account possible new risks arising from the risk control process. If residual risk is judged as nonacceptable, a new risk control procedure is performed.

Risk/Benefit Analysis. If further risk control is not practicable, the residual risk has to be weighted against the medical benefit of the device's intended use.

IV. *Overall Risk Acceptability and Final Report.* The acceptability of the overall residual risk has to be discussed and defined. A risk management report reviews the process, the data, and the decision of the acceptability of the overall residual risk.

V. *Postmarket Surveillance.* A procedure has to be implemented for gathering and evaluating the postmarket information, so as to judge whether elements of the risk management process, and especially some risk estimations, need to be revised and updated.

The standard makes available also a wide rationale for its requirements (Annex A), a checklist for identifying characteristics, which could impact on safety (Annex C), and a discussion of main risk concepts for the medical devices (Annex D).

7.9.2.4   Key Areas

- *Outsourcing.* Outsourced processes or activities and internal ones, even if residuals, have to be incorporated to assure a full risk management.
- *Design and Development.* Integral of these activities are the processes of risk identification and control. Then, the risk elements should participate in design input, output, verification, and validation.
- *Traceability.* The risk situation should be used to define the traceability requirement (depth, contact, etc.) for the device.
- *Production and Servicing.* Purchases and their control, process and their control, equipment, processes validation, and servicing activities should be connected and designed taking into account the risk management outputs.
- *Corrective and Preventive Actions.* Managing actions to correct and prevent the effects of nonconformities in design and manufacturing, and to react to service reports and customer/market complaints, should be integrated as to make visible unknown problems from which new risks could arise to be studied in the risk management process.

7.9.3   Essentials of the Nonclinical Evaluation

7.9.3.1   Key Terms

- *Safety.* Freedom from *nonacceptable* risks (not from all risks).
- *Biocompatibility.* An *appropriate* tissue or biological systems response when the device is in contact with the body as foreseen in its intended use. "Appropriate" has to do with the idea of a *tolerable* response, if weighted with the medical benefits.
- *Side Effect.* The *unwanted* effects of a device or material, but inherent in it.

7.9.3.2 An Overall Evaluation The nonclinical evaluation contains everything that concerns the device safety for patient and operators, different from the clinical performance and safety. In general, there are a lot of aspects that can impact *overall* (patient, professional user, operators, population) safety:

- *Energy*: electricity, heat, mechanical forces, radiations, motion, etc.
- *Environmental*: electromagnetic, power supply, connection or coordination with other devices, accidental damage, incorrect storage and transport, waste disposal, etc.
- *Outputs*: energy, medical agents, pressure, etc.
- *Software/firmware*: failure, errors, validation, etc.
- *Failures*: accidental damages, inadequate maintenance, aging, end-of-life, loss of integrity, etc.
- *Use and user*: labeling, instructions adequacy and understandability, information on side effects, incompatibility with materials, substance, or other devices, human error, etc.
- *Man-device interface*: usability, readability, nonambiguity, controls, etc.
- *Biological*: biocontamination, incorrect formulation, biological toxicity (systemic, local, allergic), genotoxicity, immunotoxicity, hygienic safety, degradation, etc.
- *Sterilization*: status, procedures, single-use, etc.
- *Nonchemically mediated hazards*: powders inhalation, eyes irritation, etc.
- *Animal testing* for performance and safety.

The most complex section can be the biological evaluation process, which requires, in the absence of well-developed literature, the direction of a competent, and well-trained expert in planning the data supply (literature, tests), in planning and evaluating the possible tests, and in assessing a multiplicity of data, often within some range.

7.9.3.3 International Standards The ISO standards give to the nonclinical biological evaluation framework a fundamental contribution, making available guidelines, principles, and detailed test procedures. The leading documents are the *ISO 10993-1:2009 Biological evaluation of medical devices. Part 1: Evaluation and testing within a risk management process*, together with its guidance paper *ISO/TR 15499:2012 Biological evaluation of medical devices. Guidance on the conduct of biological evaluation within a risk management process.* (*ISO/TR* specifies a "technical report", an informative document of a different kind from that of normative standards, not having the status of a standard). The ISO 10993 series then has a widely comprehensive list of titles:

- **ISO 10993-2:2006** Biological evaluation of medical devices—Part 2: Animal welfare requirements.

- **ISO 10993-3:2003** Biological evaluation of medical devices—Part 3: Tests for genotoxicity, carcinogenicity, and reproductive toxicity (as of today, it is under revision).
- **ISO 10993-4:2002** Biological evaluation of medical devices—Part 4: Selection of tests for interactions with blood.
- **ISO 10993-5:2009** Biological evaluation of medical devices—Part 5: Tests for *in vitro* cytotoxicity.
- **ISO 10993-6:2007** Biological evaluation of medical devices—Part 6: Tests for local effects after implantation (as of today, it is under revision).
- **ISO 10993-7:2008** Biological evaluation of medical devices—Part 7: Ethylene oxide sterilization residuals.
- **ISO 10993-9:2009** Biological evaluation of medical devices—Part 9: Framework for identification and quantification of potential degradation products.
- **ISO 10993-10:2010** Biological evaluation of medical devices—Part 10: Tests for irritation and skin sensitization.
- **ISO 10993-11:2006** Biological evaluation of medical devices—Part 11: Tests for systemic toxicity.
- **ISO 10993-12:2012** Biological evaluation of medical devices—Part 12: Sample preparation and reference materials.
- **ISO 10993-13:2010** Biological evaluation of medical devices—Part 13: Identification and quantification of degradation products from polymeric medical devices.
- **ISO 10993-14:2001** Biological evaluation of medical devices—Part 14: Identification and quantification of degradation products from ceramics.
- **ISO 10993-15:2000** Biological evaluation of medical devices—Part 15: Identification and quantification of degradation products from metals and alloys.
- **ISO 10993-16:2010** Biological evaluation of medical devices—Part 16: Toxicokinetic study design for degradation products and leachables.
- **ISO 10993-17:2002** Biological evaluation of medical devices—Part 17: Establishment of allowable limits for leachable substances.
- **ISO 10993-18**:2005 Biological evaluation of medical devices—Part 18: Chemical characterization of materials.
- **ISO/TS 10993-19:2006** Biological evaluation of medical devices—Part 19: Physicochemical, morphological and topographical characterization of materials.
- **ISO/TS 10993-20:2006** Biological evaluation of medical devices—Part 20: Principles and methods for immunotoxicology testing of medical devices.

(ISO/TS specifies a "technical specification," normative document representing the technical consensus within a group of ISO experts.)

The nonclinical evaluation always requires to be intended in the context of a full risk management process, which is fundamental in identifying the nonbiological components of the evaluation. In this way, the biological competences go together with the competences of the manufacturer, gained from the experience of comparable or competitive devices, from the market experiences, and from the servicing and maintenance records, and frequently available in the widest technical and scientific environment.

7.9.4   Essentials of the Clinical Evaluation *(Source:* IMDRF Documents GHTF/SG5/N1R8:2007-N2R28:2007-N3:2010)

7.9.4.1   Key Terms

- *Clinical investigation.* A systematic study in human subjects, undertaken to assess the safety or performance of a medical device.
- *Clinical data.* A safety/performance information generated from the clinical use of a medical device.
- *Clinical evaluation.* The assessment and analysis of clinical data to verify the safety and performance of a medical device.
- *Clinical evidence.* The clinical data and the clinical evaluation report for a medical device.

7.9.4.2   General Principles for Assessing a Clinical Investigation Need   A clinical investigation is necessary to provide the data not available from literary sources. For well-established products, materials, and technologies, a clinical investigation may not be necessary and the available clinical data should be sufficient.

To specify the need of an investigation

- relevant clinical essential principles should be identified;
- a risk management procedure should help in addressing residual risk or performance aspect not yet resolved or coming from new claims or new uses;
- a clinical evaluation procedure should demonstrate missing clinical data and the possible contribution of available sources.

7.9.4.3   General Principles for Designing a Clinical Investigation   Any clinical investigation must be based on the result of a clinical evaluation process and follow a proper risk management procedure to avoid undue risks.

Factors that may influence the clinical data requirements should be

- device type and regulatory classification;
- new technology against previous experience;
- clinical indications and performance claims by the manufacturer;
- exposure (surface contact, invasive, etc.) and period of exposure;
- component materials and substances;

- risk inherent in the use of the product/procedure;
- lifetime of the device and potential impact of the device failure;
- disease process and patient population being treated; cultural and demographic considerations (age, ethnicity, etc.);
- availability of alternative treatments and current standards;
- ethical considerations.

Factors that should be considered in the study design should include

- statement of clear objectives;
- evaluation of an appropriate subject population;
- minimization of bias and analysis of possible confounding factors;
- appropriate controls (cohort, historical, sham, etc.);
- design configuration (parallel, crossover, etc.);
- comparison types (superiority, equivalence, etc.);
- statistical considerations, methods, and techniques.

The outcome of a clinical investigation should be documented in a final study report, which forms a part of the wide clinical data employed within the clinical evaluation process.

7.9.4.4 *International Standards*

A detailed guide to the *Clinical Investigation* is given by the Standard *ISO 14155:2011 Clinical investigation of medical devices for human subjects—Good clinical practice.*

The standard declares as scope the protection of the rights, safety, and well-being of human subjects, the assurance of the scientific conduct of the investigation and the credibility of its results and the definition of the responsibilities of the investigation actors. The activities covered by the standard are

- the ethical questions: accordance with the Helsinki Declaration, relations with the Ethics Committees, informed consent and its procedures, etc.;
- the clinical investigation plan: risk evaluation, justification for the need and the planning, information, case reporting, monitoring, etc.;
- the clinical investigation process: site inspection and monitoring, adverse effects and device deficiencies, documents and data/documents approval, traceability, control, investigation auditing, etc;
- end of the investigation: suspension, resuming, close-out, reporting;
- the responsibilities of the organization and the investigator(s).

The standard offers a detailed guideline for studying and writing down a Clinical Investigation Plan (Annex A), for preparing a Case Report Form (Annex C), and for writing down the Clinical Investigation Report (Annex D).

7.9.4.5   The General Process of a Clinical Evaluation

Identification and Collection of Clinical Data. Data can be found at the manufacturer or in the scientific literature. The manufacturer is responsible for identifying relevant data. The data can come from any of following:

- Literature searching, where a protocol for identifying the sources, the database and its extent, the selection criteria adopted, and the strategies for avoiding data duplication should exist. The literature search should generate a report addressing the search protocol, the search report and the article, or other references.

- Clinical experience, where postmarket surveillance reports, adverse effects databases, and relevant corrective actions by the manufacturer should be considered. The clinical experience analysis demands deep and complete information on every case studied: it provides a "real-world experience," often mediated by less experienced actors.

- Clinical investigation(s), where the clinical investigation plan and its amendments, the case report forms, the records of monitoring and auditing, the Ethics Committee and Regulatory Authority documents and approvals, and the final report should be available.

Appraisal of Clinical Data. Data should be assessed for the search and providing strategy and credibility, and for quality, acceptability and relevance to the device, and its intended use. Data should be evaluated following a set of criteria, particularly in examining the effectiveness of the clinical study strategy. The clinical data, where possible, should be weighted to assess the strength of the data set.

Analysis. The scope of the analysis is to determine whether the appraised data set is sufficient and acceptable for demonstrating the clinical performance and safety of the device. If not, an incremental stage of identification and collection of data has to be performed. The set of pivotal data can be identified, validated, and compared. The analysis should be able to show that

- the device performs as intended;

- the device does not pose any undue safety concerns and that any residual risk is acceptable when weighted against the benefits;

- the device literature (information, instruction, labels, etc.) is consistent with the data collected.

Reporting. The clinical evaluation report should be a stand-alone document, containing sufficient information for the reading by a third party. The detail level will depend on the scope of the clinical evaluation. The document should address the following points:

(a) device identification,

(b) device description (materials, components, characteristics, etc.), intended ways of use (sterile/nonsterile, invasive/noninvasive, etc.), and classification,

(c) intended therapeutic indications and claims,

(d) device context (technology history and advancement, comparable devices, manufacturing facts, etc.), relevant essential principles, and connected choice of clinical data,

(e) summary of clinical data and their appraisal,

(f) data analysis and data set used and their relative importance for evaluating the *performance*, the *safety*, and the *product literature*,

(g) conclusion, clearly stating, for every proposed clinical indication, whether
  • the clinical evidence demonstrates conformity with the relevant essential principles;
  • the device performs is as intended and claimed;
  • the device safety is as intended and claimed;
  • any residual risk is acceptable when weighted against the benefits;
  • the device literature (information, instruction, labels, etc.) is correct and sufficient.

REFERENCES

[1] Kramer DB, Xu S, Kesselheim AS. Regulation of medical devices in the United States and European Union. N Engl J Med 2012;366:848–855.

[2] European Commission. Medical devices. Available at http://ec.europa.eu/consumers/sectors/medical-devices/. Accessed 2013 July 11.

[3] European Commission, Public Health, Revision of the Medical Device Directives, Website [accessed on 2013 July 9]: http://ec.europa.eu/health/medicaldevices/documents/revision/.

[4] Zuckerman DM, Brown P, Nissen SE. Medical device recalls and the FDA approval process. Arch Intern Med 2011;171(11):1006–1011.

[5] Medical Devices – Overview of Device Regulation. Website [accessed on 2013 July 11]: http://www.fda.gov/MedicalDevices/DeviceRegulationandGuidance/Overview/default.htm.

[6] Principles of conformity assessment for medical devices SG1 final document GHTF/SG1/N78:2012; 2012 Nov 2, http://www.imdrf.org/docs/ghtf/final/sg1/technical-docs/ghtf-sg1-n78-2012-conformity-assessment-medical-devices-121102.pdf (Accessed on April 20, 2016).

[7] Yamaguchi U, Chuman H. Overview of medical device regulation in Japan as it relates to orthopedic devices. J Orthop Sci 2013;18(5):866–868.

[8] Regulatory System for Medical Devices in Japan, Pharmaceuticals and Medical Devices Agency, Standard for medical devices. Website [accessed on 2013 July 11]: http://www.std.pmda.go.jp/stdDB/index_e.html.

[9] Australian Regulatory Guidelines for Medical Devices V1.1 May 2011, Department of Health and Ageing – Therapeutic Goods Administration. https://www.tga.gov.au/sites/default/files/devices-argmd-01.pdf (Accessed on April 20, 2016).

[10] Kambeitz T. Medical Device Approvals in Brazil: A Review and Update. Underwriters Laboratories Inc.; 2011. http://library.ul.com/wp-content/uploads/sites/40/2015/02/UL_WP_Final_Medical-Device-Approvals-in-Brazil_v7_HR.pdf (Accessed on June 6, 2016).

[11] Machado AFP, Hochman B, Tacani PM, Liebano RE, Ferreira LM. Medical devices registration by ANVISA (Agencia Nacional de Vigilancia Sanitaria). Clinics 2011;66(6):1095–1096.

[12] Consolidation Medical Devices Regulations SOR/98-282, May 26, 2013-Last amended on December 16, 2011. Website [accessed on 2013 June 19]: http://laws-lois.justice.gc.ca.

[13] Medical Devices Chapter 2: Canadian Requirements. Website [accessed on 2013 July 17]: http://www.hc-sc.gc.ca/dhp-mps/mdim/activit/fsfi/meddevfs_matmedfd-eng.php.

[14] Lamph S. Regulation of medical devices outside the European Union. J R Soc Med 2012;105:S12–S21. DOI: 10.1258/jrsm.2012.120037.

[15] Medical Device Regulations. *Global Overview and Guiding Principles*. WHO; 2003.

[16] Medical Devices Regulation: Website [accessed on 2013 July 17]: http://www.globalregulatorypress.com/.

INDEX

Biomedical Devices: Design, Prototyping, and Manufacturing, First Edition.
Edited by Tuğrul Özel, Paolo Jorge Bártolo, Elisabetta Ceretti, Joaquim De Ciurana Gay,
Ciro Angel Rodriguez, and Jorge Vicente Lopes Da Silva.
© 2017 John Wiley & Sons, Inc. Published 2017 by John Wiley & Sons, Inc.